Exercise, Sport, and Bioanalytical Chemistry

Emerging Issues in Analytical Chemistry

Series Editor
Brian F. Thomas

AMSTERDAM · BOSTON · HEIDELBERG · LONDON
NEW YORK · OXFORD · PARIS · SAN DIEGO
SAN FRANCISCO · SINGAPORE · SYDNEY · TOKYO

Exercise, Sport, and Bioanalytical Chemistry
Principles and Practice

Anthony C. Hackney, PhD, DSc
Department of Exercise & Sport Science,
Department of Nutrition, University of North Carolina,
Chapel Hill, NC, United States

AMSTERDAM • BOSTON • HEIDELBERG • LONDON
NEW YORK • OXFORD • PARIS • SAN DIEGO
SAN FRANCISCO • SINGAPORE • SYDNEY • TOKYO

Elsevier
Radarweg 29, PO Box 211, 1000 AE Amsterdam, Netherlands
The Boulevard, Langford Lane, Kidlington, Oxford OX5 1GB, UK
50 Hampshire Street, 5th Floor, Cambridge, MA 02139, USA

Published in cooperation with RTI Press at RTI International, an independent, nonprofit
research institute that provides research, development, and technical services to government
and commercial clients worldwide (www.rti.org). RTI Press is RTI's open-access, peer-reviewed
publishing channel. RTI International is a trade name of Research Triangle Institute.

Library of Congress Cataloging-in-Publication Data
A catalog record for this book is available from the Library of Congress

British Library Cataloguing-in-Publication Data
A catalogue record for this book is available from the British Library

ISBN: 978-0-12-809206-4

For Information on all Elsevier publications
visit our website at http://www.elsevier.com/

Working together
to grow libraries in
developing countries

www.elsevier.com • www.bookaid.org

Typeset by MPS Limited, Chennai, India

DEDICATION

To my wonderful family—Sarah, Zachary, and Grace. Thank you for all your support.

CONTENTS

FOREWORD

Professor Anthony Hackney—"Tony"—and I received our doctorate degrees in the same year (1986), but our careers started more than 7000 km apart, mine in Finland, and his in the United States. Over the past 30 years we have watched each other's research agenda grow and develop. At times our research interests have overlapped extensively and other times they have diverged greatly. Regardless of the focus of our research, he and I have had a great mutual respect for one another and the work our research groups have been doing. Throughout his career he and his students have produced hundreds of research articles and book chapters in print as well as numerous national and international presentations. Tony's curiosity, drive, and desire to understand the workings of the exercising human are evidenced by this scholarly productivity and the passion he displays for his work.

Tony is well known internationally for his ongoing pursuits in exercise endocrinology in which he is one of the most eminent researchers in the world. This is apparent from his varied and numerous international faculty appointments, research fellowships, as well as his service work in 40 countries—he is truly a "world citizen." For a scientist to study the world of hormones successfully, as Tony has, it takes a meticulous approach, attention to detail, and a keen analytical mind. Tony has brought all these characteristics to this book. In addition, he has clearly and concisely presented and explained the difficult scientific concepts associated with exercise biochemistry, endocrinology, and physiology—his goal is for people to understand the complexities of this topic. He has been exceedingly successful in his endeavors. I congratulate him on this project and I encourage the readers to immerse themselves, and enjoy.

Keijo Häkkinen PhD
Department of Biology of Physical Activity,
Faculty of Sport and Health Sciences, University of Jyväskylä,
Jyväskylä, Finland

PREFACE

I believe we have few clear epiphanies in life, but I can think back now and recognize one. In 1977 I was in Freiburg, West Germany (Federal Republic of Germany) as a visiting student. I had just heard a lecture at the university there on sports medicine and how science could be used to develop human performance. I had always participated in sports, though not very well, and wondered why some people's athletic ability responded so well to exercise training and others' did not. This lecture opened my eyes to the potential for systematic scientific study of the issue. I was hooked. When I returned to my school, Berea College in the United States, I devoted myself to gaining the education and experience necessary to become a sports scientist who tries to answer the question "How does the body work in exercise and how do we make it work better?" This has been my professional passion and goal. Four decades later, I have slowed down a little physically, but I still have a passionate desire to try and fully understand that "how" question. That passion is why I cannot wait to get to work most days and start on new projects, why I enjoy so much working with the students at my university, and why I wanted to write this book.

The book's primary objective is to discuss the biochemistry and bioanalytical techniques used to understand the physiological processes, assessment, and quantification of physical activity, exercise, and sport. A secondary objective is to describe procedures and practices for improving the capacity to perform exercise, which can lead to improved health and sports performance.

The terms "physical activity" and "exercise" are often used interchangeably, but they have different technical definitions. Everyday action that requires muscular contraction (walking to the mailbox, mowing the grass, washing dishes) is physical activity. Exercise is any deliberate physical activity (jogging, weightlifting, playing basketball) done with the purpose of improving health, fitness, or sporting performance. Thus, *all exercise is physical activity, but not all physical activity is exercise.* "Sport" is sometimes used in the same context as physical activity and exercise, and it has overlapping aspects with these terms, but it too has a technically distinct definition. Sport is an athletic event requiring some skill and physical ability and often has a competitive element; within the discussions herein it will be delimited by having a high degree of physical activity as a key component. These three terms, while different in definition, do have a commonality: the biochemical and physiological responses to physical activity, exercise, and sport all vary as a function of the physical stress placed upon the body—the greater the level of stress, the greater the responsiveness. For simplicity, in this book the general term referring to all three will be "exercise," and "physical activity" and "sport" will only be used when specificity is desirable.

Why this book, when so many others exist? Many of the available books are written at one of two levels: more or less simplistically for the general public or technically for the specialist with the requisite formal education and knowledge. This book attempts to occupy the middle ground by offering the fundamental biochemistry and some elements of the physiology behind exercise and describing the analytical methods used to understand it. It will inform the specialist of emerging knowledge, trends, and techniques, and allow the nonspecialist to grasp the underlying science and current practice of the discipline relatively quickly.

ACH

ACKNOWLEDGMENTS

I need to express sincere thanks to my graduate students at the University of North Carolina for their patience in working with me throughout this project.

I must also thank my colleagues at Tartu University in Estonia, especially Drs Vahur Ööpik and Mehis Viru, for their valuable insight and guidance.

I greatly appreciate the help of my daughter Sarah, who has a keen eye for detail and a tremendous way with words.

Most certainly I must acknowledge the great help and support of the people at RTI International in North Carolina, Drs Gerald T. Pollard and Brian F. Thomas, who really made this all possible. Also, my thanks to Dayle G. Johnson of RTI International for the cover design.

PART *I*

Introduction: Basics and Background

CHAPTER *1*

Energy and Energy Metabolism

The aim of this chapter is to provide an overview of energy forms and types, and how chemical energy formation through biochemical reactions is essential for physiological functions in the healthy human body at rest and during the active state of exercise.

ENERGY

In physics, energy is a property that can be transferred between objects or states. The ability of a system to perform work is a more biological definition of energy and a meaning that can be applied to humans. Work in the biochemical and physiological sense is referred to as the energy transferred by mechanical means, or simply force applied or acting over a distance. During exercise, muscle does the work; that is, it applies a force (which requires energy) over a selected movement, distance, and pattern.[1]

ENERGY TRANSFORMATION

The first law of thermodynamics, also called the conservation of energy principle, states that energy can be neither created nor destroyed, but it can exist in different classification forms: chemical, thermal, nuclear, electromagnetic, electrical, and mechanical. The body uses the first law daily as it consumes food, a form of chemical energy (macronutrients; see Chapter 2, "Energy Metabolism of Macronutrients During Exercise"), and converts it to useful chemical energy in the form of adenosine triphosphate (ATP).[2] ATP consists of a base substance, adenine, attached to a sugar, the carbohydrate ribose, which has three phosphate molecules attached by high energy bonds. Removal or breakage of these phosphate bonds provides the energy for bodily processes.

CHEMICAL ENERGY OF THE BODY

The body is so dependent on ATP that it is called the "energy currency." All physiological processes require it. ATP dependency is especially true for skeletal muscle, which provides the movement during exercise. The biochemical reaction by which ATP delivers useful energy can be represented as follows:

$$ATP \rightarrow ADP + Pi + Energy$$

This reaction liberating energy involves the biochemical process of hydrolysis. ATP is broken down into adenosine diphosphate (ADP) by breakage of one of the high energy bonds and removal of a phosphate (Pi). The process is reversible; that is, rephosphorylation of ADP to ATP can occur by reattaching a Pi using the energy contained in food macronutrients when they are metabolized in select biochemical pathways.[3] These ATP synthesis pathways are explained in more detail later but are simplistically categorized biochemically as either anaerobic (not requiring oxygen) or aerobic (requiring oxygen). Many exercise activities rely predominantly on one pathway. All-out explosive muscular movements such as sprinting 100 meters (m) as hard and fast as possible is primarily anaerobic and requires provision of ATP rapidly over a short period. A 20-kilometer (km;1 km = 0.62 mile) run is predominantly aerobic; the muscular movement is not nearly as rapid, hence ATP can be produced more slowly, but large quantities are needed.[4,5]

ENZYMES

Biochemical reactions are catalyzed—that is, regulated—by proteins called enzymes. Enzymes influence the speed of a reaction, affecting the energy of activation, but they do nothing to alter the outcome. In the simple reaction formula below, the formation of the chemical products C and D could be occurring 100 million times quicker in the presence of an enzyme.[6]

$$A + B \xrightarrow{\text{Enzyme}} C + D$$

Chemical Reactants Chemical Products (Substrates)

The ATP reaction above is regulated in skeletal muscles as follows:

$$ATP \xrightarrow{\text{Actomyosin ATPase (enzyme)}} ADP + Pi + Energy$$

$$ADP + Pi + Energy \xrightarrow[\text{ATP Synthase (enzyme)}]{} ATP$$

At the cellular level, signaling agents such as hormones can affect the activity rate of enzymes and in turn influence the speed at which reactions such as hydrolysis and rephosphorylation proceed (see Chapter 3, "Regulation of Energy Metabolism During Exercise"). These agents are among the major ways in which physiological events occurring at the cellular level are regulated.[7]

ENERGY CONSUMPTION

ATP is measured in moles (mol), but when researchers quantitate energy in humans it is typically done in kilocalories (kcal) per mol in the United States or kilojoules (kJ) per mol in Europe. The energy released by 1 mol of ATP is approximately 7.3 kcal or 30.5 kJ.[8] The kcal (sometimes called a Calorie in the United States) is actually a thermal unit of energy developed over a century ago and represents the amount of heat energy necessary to raise the temperature of 1 kilogram (kg) of water 1°C. It can be used to express the amount of chemical energy contained in food items as well as energy liberated when exercise is performed (first law of thermodynamics).

ENERGY TRANSFORMATION IN EXERCISE

The average adult human (male 70 kg, female 62 kg) expends about 1 kcal/min (males slightly more, females slightly less) in a resting state. This is the resting metabolic rate (RMR), the amount of energy necessary each day to "just exist." The RMR term is sometimes used interchangeably with basal metabolic rate (BMR), but the two are not exactly the same; BMR is a more rigorously controlled scientific measurement (see Chapter 5, "Energy Expenditure at Rest and During Various Types of Physical Activity," and Chapter 6, "Energy Storage, Expenditure, and Utilization: Components and Influencing Factors").

As an illustration of energy transformation, assume you have a representative candy bar which contains about 250 kcal. Theoretically, if you ate it and remained in the resting state, you would expend that energy in a little over 4 hours (h). By contrast, if you went for a 5 km jog at a moderate pace, you would expend it

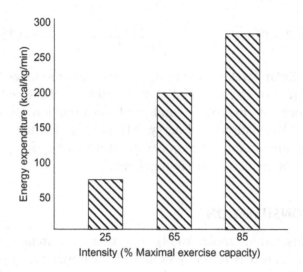

Figure 1.1 Influence of the intensity of an activity on the rate of energy expenditure (kcal) per kg of body mass per min.[10]

Table 1.1 Examples of the Influence of the Duration of an Activity on the Total Energy Expended if Walking 5 km in 1 h[9]			
Mass (kg)	Energy Expenditure Rate (kcal/min)	Time (min)	Total Energy Expended (kcal)
62	3.5	15	52.5
		30	105
		60	210
70	4.0	15	60
		30	120
		60	240

in about 30 minutes (min).[9] To expend that amount of energy when jogging in one-fourth the time, the energy producing biochemical pathways have to speed up the process by which ADP gets rephosphorylated into ATP. These accelerated energy expenditure and production rates, commonly referred to as burning energy, rely on converting the chemical energy in the macronutrients of food into ATP more rapidly. The rate at which energy is burned during exercise is a function of the intensity of the muscular work being done, and the total energy burned is a function of the duration of exercise (Fig. 1.1 and Table 1.1).

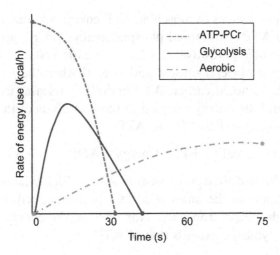

Figure 1.2 The predominant energy pathways used when performing maximal exercise over a varying amount of time.

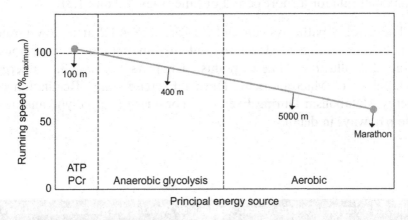

Figure 1.3 The predominant energy pathways used when running at maximal effort over different distances (marathon = 42.2 km).

Similarly, the major biochemical pathways for ATP production are influenced by and dependent upon the intensity and duration of the activity. Energy is always being derived through both anaerobic and aerobic pathways, but, depending on the activity, one of them almost always predominates. What occurs as the body shifts from the predominance of one pathway to the other is the energy continuum, which is illustrated in practical terms by Figs. 1.2 and 1.3.

The principal sources of muscular ATP energy are anaerobic hydrolysis of stored ATP (plus stored phosphocreatine [PCr]), anaerobic glycolysis, and aerobic pathways, of which there are several. ATP–PCr is stored directly in skeletal muscle and is used when there is a need to increase ATP. As noted earlier, ATP hydrolysis releases useful energy; PCr is split, and the energy released in the hydrolysis of the phosphate is used to rephosphorylate ADP to ATP:

$$PCr \rightarrow Creatine + Pi + Energy \rightarrow ADP + Pi \rightarrow ATP$$

Stores are limited, however, because ATP is highly labile, and there are physical limits to the amount of PCr that tissue will hold, in part due to its hydrophilic properties. Nonetheless, the energy quantity of muscular PCr typically exceeds that of ATP.

The anaerobic glycolytic pathway is a rapid energy source, but the amount of ATP that can be produced is limited and can only serve at a maximal rate for a short period of time (Figs. 1.2 and 1.3).

The aerobic pathways, in contrast, produce ATP at a slower rate than PCr or anaerobic glycolysis, but the amount can be enormous. Table 1.2 illustrates these points. Chapters 2 and 3, "Energy Metabolism of Macronutrients During Exercise" and "Regulation of Energy Metabolism During Exercise," consider the glycolytic and aerobic pathways in detail.

Table 1.2 Example of Energy Source Available to Working Muscle, Assuming 70 kg Body Mass and Average Body Composition[4]					
	Energy Source				
	ATP	PCr	Anaerobic Glycolysis	Aerobic (Carbohydrate)	Aerobic (Lipid)
Energy amount (g)	40	120	350	500	15,000
Duration until depletion (time)	4–6 s	15–20 s	1–2 min	1–2 h	>6 h
Maximum rate of synthesis (mmol/kg/s)		9	4	2	1
mmol/kg/s, millimole ATP per kilogram body mass per second.					

Close-Up: Technological Advances in Measuring Chemical Energy in Humans: From Muscle Biopsy to Magnetic Resonance Imaging

How do you measure the amount of ATP in a human body? The bioanalytical process for in vivo (Latin for "within the living") determination has changed over the years. An early procedure was the needle biopsy, the surgical extraction of a very small piece of muscle. The sample was chemically stained to allow microscopic determination of morphological components and biochemically analyzed to determine chemical constituents such as ATP. The classical technique used in most exercise studies is the Bergstrom procedure, named after the Swedish scientist who popularized it in the 1960s.[11] For much of the 20th century, this invasive method was the gold standard for quantification of ATP.

Over the past 40 years, nuclear magnetic resonance spectroscopy of β-phosphorus atoms (^{31}Pβ-NMR) became the preeminent technique for determining the structure of organic compounds such as ATP. The absorption and emission of energy from nuclei in a magnetic field are recorded, so the procedure does not require removal of tissue from the body. All that is required is placement of a body segment inside a radio frequency coil device that transmits and receives signals. A variety of names and abbreviations have been used to refer to the process: in the 1940s, nuclear induction; in the early 1950s, nuclear paramagnetic resonance; since the late 1950s, nuclear magnetic resonance.

Because of patients' concerns about nuclear energy, radioactivity, and the like, by mid-1980s the use of the term nuclear had been largely eliminated and replaced by just magnetic resonance (MR) imaging or MRI. The lexicon has further expanded to include MR angiography (MRA), MR spectroscopy (MRS), and functional magnetic resonance imaging (fMRI). Interestingly, for uncertain reasons, most scientific journals prefer MR imaging to MRI.

Now assessment of ATP is painless and noninvasive. But the new techniques are very expensive and can be somewhat nonspecific for isolating events at the single cell level of function. For these reasons, you may still see needle biopsy reported in contemporary research literature even though the procedure is over 50 years old.

REFERENCES

1. Jammer M. *Concepts of Force*. Cambridge, MA: Harvard University Press; 1957.

2. Kamerlin SC, Warshel A. On the energetics of ATP hydrolysis in solution. *J Phys Chem B*. 2009;113(47):15692–15698.

3. Nelson DL, Cox MM. *Lehninger Principles of Biochemistry*. 6th ed. New York: Macmillan Higher Education; 2013.

4. Gastin PB. Energy system interaction and relative contribution during maximal exercise. *Sports Med*. 2001;31(10):725–741.

5. Joyner MJ, Coyle EF. Endurance exercise performance: the physiology of champions. *J Physiol*. 2008;586(Pt 1):35–44.

6. Suzuki H. *Chapter 8: Control of enzyme activity. How Enzymes Work: From Structure to Function*. Boca Raton, FL: CRC Press; 2015:141–169.

7. Hackney AC. Stress and the neuroendocrine system: the role of exercise as a stressor and modifier of stress. *Expert Rev Endocrinol Metab*. 2006;1(6):783–792.

8. Hargrove JL. Does the history of food energy units suggest a solution to "Calorie confusion"? *Nutr J*. 2007;6(44). Available from: <http://dx.doi.org/10.1186/1475-2891-6-44>.

9. America College of Sports Medicine. *ACSM's Resource Manual for Guidelines for Exercise Testing and Prescription*. 7th ed. New York: Lippincott Williams & Wilkins; 2013.

10. Romijn JA, Sidossis LS, Castaldelli A, Horowitz JF, Endert E, Wolfe RR. Regulation of endogenous fat and carbohydrate metabolism in relation to exercise intensity and duration. *Am J Physiol*. 1993;265(3):E380–391.

11. Bergstrom J. Muscle electrolyte in man. Determined by neutron activation analysis on needle biopsy specimens. *Scand J Clin Lab Invest*. 1962;14(suppl 68).

CHAPTER 2

Energy Metabolism of Macronutrients During Exercise

The aim of this chapter is to provide an overview of the biochemical pathways by which dietary macronutrients—carbohydrates, fats, and proteins—are turned into useable chemical energy in the form of adenosine triphosphate (ATP) by the human body.

OVERVIEW OF METABOLIC ENERGY PATHWAYS

In biochemistry, metabolism is defined as the chemical processes that occur within a living organism in order to maintain life. There are many aspects to metabolism, but in this chapter the discussion is limited to energy metabolism, that is, how the chemical energy in food is converted to ATP. As introduced in Chapter 1, "Energy and Energy Metabolism," the biochemical pathways associated with the conversion of food chemical energy into ATP are classified as either anaerobic or aerobic. Fig. 2.1 shows which energy systems and pathways fit into the anaerobic and aerobic classifications as related to skeletal muscle energy production.

CARBOHYDRATE METABOLIC PATHWAYS

Nutrients into Glucose

The typical Western diet is comprised of approximately 50% carbohydrate, 15% protein, and 35% fat.[1] Carbohydrates provide the major source of food chemical energy that can be converted to ATP, and energy pathways are structured so that carbohydrate metabolism is a major crux for the production of ATP. Fig. 2.2 illustrate this point, showing that noncarbohydrate macronutrients also enter into elements of carbohydrate biochemical pathways to ultimately yield ATP. Noncarbohydrate energy production is discussed later in the chapter.

The carbohydrates in food consist of simple (monosaccharide and disaccharide, sometimes called simple sugars) and complex

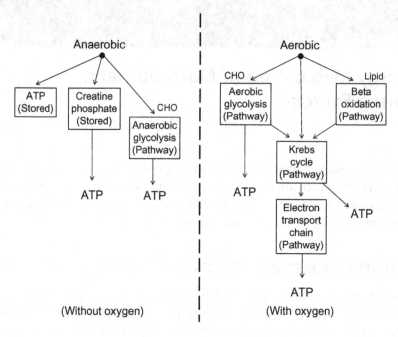

Figure 2.1 *Classification of energy metabolism pathways as either anaerobic or aerobic.*

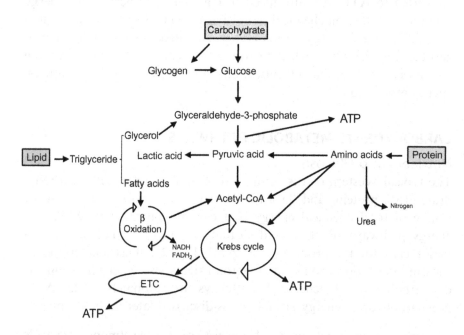

Figure 2.2 *How the macronutrients enter into the major energy metabolism pathways resulting in ATP production.* ETC, *electron transport chain;* Lipids, *fats.*

(polysaccharide) forms. Regardless of the form ingested and digested, the vast majority of the energy metabolism of carbohydrate revolves around the monosaccharide glucose ($C_6H_{12}O_6$), which is one of the most plentiful and common dietary sugars.[1,2] In common parlance, blood sugar means blood glucose.

Glucose has several biochemical mechanisms by which it or its breakdown products interact in pathways to result in the rephosphorylation of ADP to ATP and provide useful energy. These pathways are intricate and have many chemical reactions. Fig. 2.2 shows the basic elements and give a simplistic overview of the means by which glucose yields ATP.

Glycolysis, the Krebs Cycle, and the Electron Transport Chain

Glucose enters the glycolytic pathway to yield ATP, and this pathway can be either anaerobic or aerobic form. The ATP yield with anaerobic glycolysis is less than aerobic, but the production of ATP is extremely rapid and independent of oxygen requirements. The anaerobic form results in the production of lactic acid (which releases an H^+ ion and becomes lactate) as an end product. Lactate is sometimes viewed as a "bad" byproduct of anaerobic metabolism, but this is a misnomer. Lactate production is essential to allow the anaerobic pathway to proceed. Furthermore, lactate is removed from skeletal muscle and placed in the blood where the liver can clear it and use it to remake glucose in a process called the Cori cycle.[2]

The aerobic form of glycolysis has a higher total ATP yield, but the rate (speed) is slower in part due to the oxygen requirement. The end product of aerobic glycolysis is acetyl coenzyme A (acetyl-CoA). Acetyl-CoA enters the Krebs cycle where it is used to produce more ATP. Anaerobic and aerobic glycolysis takes place in the cytosol of a cell (in skeletal muscle, referred to as sarcoplasm), while the Krebs cycle pathway takes place in the mitochondria, found extensively in skeletal muscle (Fig. 2.3). The Krebs cycle is named after Sir Hans Krebs, a German-born British biochemist who won the 1953 Nobel Prize in Physiology or Medicine for his work in understanding energy metabolism. The proper biochemistry name for the Krebs cycle is the tricarboxylic acid cycle or the citric acid cycle.[2]

The Krebs cycle is a high-yield pathway for ATP production, but little ATP is directly produced. What does get produced is large

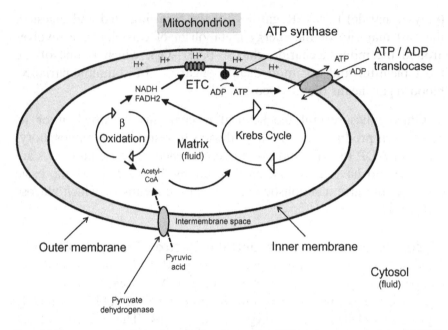

Figure 2.3 The energy metabolism pathways in the mitochondrion. ETC, electron transport chain; ATP/ADP translocase moves adenosine equivalents in and out of mitochondria.

amounts of nicotinamide adenine dinucleotide (NADH) and flavin adenine dinucleotide (FADH2), which can go through reduction−oxidation (redox) chemical reactions to yield ATP.

Redox reactions are the essential underpinning for exercise energy metabolism. The oxidation phase involves the removal of electrons from ions or other molecules; the reduction phase involves the addition of electrons to relevant structures. Redox reactions are always coupled, meaning that every time there is an oxidation there has to be a simultaneous reduction. While redox reactions involve the transfer of electrons, the most common form involves the exchange of hydrogen ions between molecules. That is, every time a hydrogen atom leaves a molecule, it goes with an electron attached. For this reason, hydrogen ions are sometimes called reducing equivalents; that is, they are equivalent to electrons.[2]

The chemical principles of redox reactions are means by which NADH and FADH2 lead to ATP production. Specifically, once NADH and FADH2 are reduced in the Krebs cycle, they react with components in the biochemical pathway called the electron transport chain (ETC). The components of the ETC are embedded in the inner mitochondrial membrane (Fig. 2.3). The enzymes of the ETC oxidize NADH and FADH2, accept hydrogens, and become reduced. Subsequently, through the biochemical process called the chemiosmotic theory, ATP is produced.[3]

The chemiosmotic theory states that, as electrons are donated from NADH and FADH2 to ETC enzyme complexes, H^+ ions are extruded from the central matrix area of the mitochondria into the inter-mitochondrial membrane space (Fig. 2.3). The net effect is an accumulation of H^+ making the intermembrane area more positive with respect to the matrix area. As a result of this ion charge difference between the matrix and the intermembrane space (ie, space between the outer and inner membrane), H^+ ions diffuse through the ATP synthase enzyme embedded in the inner mitochondrial membrane. This diffusion results in the rephosphorylation of ADP to ATP and is known as oxidative phosphorylation.

On average, NADH results in three H^+ ions and FADH2 results in two H^+ ions being extruded into the intermembrane space, yielding three and two ATP, respectively. The last reaction in the ETC results in water formation from H^+ ions and oxygen, hence the aerobic classification of the Krebs cycle which is providing the NADH and FADH2 and the ETC which uses them to make ATP.

LIPID METABOLIC PATHWAYS

Fat is the second most prevalent macronutrient in the Western diet. There is some confusion in the general public about the nature of fats and their biological role. Fats belong to a chemical classification called lipids. In this text, "fat" refers to the foodstuffs and the dietary macronutrient, "lipid" refers to the substrate used in biochemical reactions and pathways.

Once fats are ingested and digested, their key lipid constituents can be used in a variety of critical physiological processes; that is, they can be constructed into hormones, incorporated into cell membranes, and form components of neurons. These uses are a key reason that fats as macronutrients are essential in the diet. The lipids from ingested fats are also used as a chemical energy source for ATP production. The energy yield (bioenergetics) from lipids is much greater than that from carbohydrates in the form of glucose (see Chapter 1, "Energy and Energy Metabolism," Table 1.2).

The most prevalent form of lipid used for ATP production is triglycerides, either directly from the diet or from the stored form in body cells, particularly adipocytes. A triglyceride consists of a glycerol molecule with three fatty acids attached. To produce ATP, the triglyceride goes through hydrolysis, where it is broken down into these basic elements. Triglycerides can have a variety of fatty acid types bonded to the glycerol, one of the most common being palmitic acid ($C_{16}H_{32}O_2$), a fully saturated fatty acid.[2]

When the triglyceride is hydrolyzed in the adipocyte or skeletal muscle during exercise, the glycerol can enter the glycolysis pathway. The glycerol is first converted to dihydroxyacetone phosphate and then to glyceraldehyde-3-phosphate (Fig. 2.2). The three free fatty acids (eg, palmitic acid) of the triglyceride enter the beta-oxidation biochemical pathway located in the mitochondrial matrix. The end result of beta-oxidation pathway reaction is conversion of the carbons of the fatty acid into acetyl-CoA. Each palmitic acid yields eight acetyl-CoA (the number of carbons in a saturated fatty acid divided by two gives acetyl-CoA yield). Since the beta-oxidation pathway occurs in the mitochondria matrix, the acetyl-CoA produced enters the Krebs cycle and ATP is produced in the ETC. The reactions in beta-oxidation also directly generate NADH and FADH2, which are used by the ETC in redox reactions to produce additional ATP. Collectively, these steps result in an extremely high-yield ATP production, but the process is extremely oxygen dependent. Lipid energy metabolism in this way is a high yield but slow rate process for ATP production.[2]

PROTEIN METABOLISM PATHWAYS

Proteins, dietary and otherwise, are comprised of amino acids. All amino acid molecules contain carbon, oxygen, hydrogen, and nitrogen atoms as the basic components (some have additional atoms such as sulfur). The nitrogen atoms combine with hydrogen to form an amine group, hence the name amino acid.

Proteins are critical, and essentially every action in the body relies on their actions. To most people, the most familiar proteins are those of skeletal muscle. They are the contractile proteins actin and myosin (myofilaments) that allow the generation of force and movements associated with all physical activity and exercise. Proteins have a multitude of other roles and can be categorized based on their function, such as enzymatic catalysis, transport, signaling, regulation, and structural.

As noted, the typical Western adult diet is about 15% protein.[1] Dietary protein is broken down into its constituent amino acids in the gastrointestinal tract. The digested amino acids enter the free amino acid pool (FAAP; Fig. 2.4). The FAAP can be complemented not only with dietary amino acids but also those catabolized from the degradation of intracellular proteins in the normal protein turnover process. Protein turnover refers to the dynamic nature of the protein content of the body, which is in a continual state of change with new ones being made (synthesis) and old ones being broken down (degradation) all the time. The process is highly energy dependent, and in the average person as much as 20% of total daily energy expenditure can be attributed to it.[2]

The amino acid content of the FAAP is constantly changing, with additions and removals, as the body attempts to maintain all of the various proteins necessary for healthy function. The amino acids in the pool can be used as a source of energy in the form of ATP. Amino acids are not a primary source of energy, given their critical role in the various categories of physiological function noted above. But under certain circumstances such as when caloric intake of food is limited (low energy availability), amino acids can be metabolized and ATP produced.[2,4]

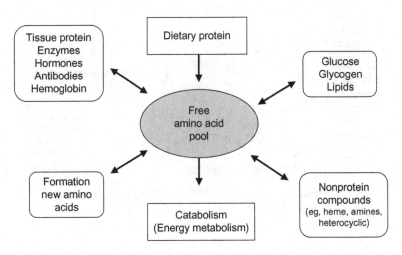

Figure 2.4 Factors affecting the free amino acid pool (FAAP).

Transamination and Oxidative Deamination

To metabolize amino acids for ATP production first involves a series of biochemical steps to remove the nitrogen-containing portion. The initial chemical reaction is transamination, where the amine nitrogen group is transferred to another substance. The resulting carbon skeleton of the amino acid without the nitrogen is converted to a variety of α-keto acids, which can be converted to reactants that enter the Krebs cycle and result in ATP formation. Fig. 2.5 shows the α-keto acid substances derived from amino acids that can enter the Krebs cycle. The process of forming Krebs cycle substances (intermediates) through transamination of amino acids is called anaplerosis.[2,4]

Transamination is reversible, providing the opportunity to build amino acids in the body, although this is limited to the nonessential amino acids. In contrast, those amino acids that cannot be built in vivo are the essential ones and must be consumed as dietary protein.

The amine nitrogen group that was initially transferred and the resulting product of transamination can proceed through another biochemical reaction, oxidative deamination. This process is not

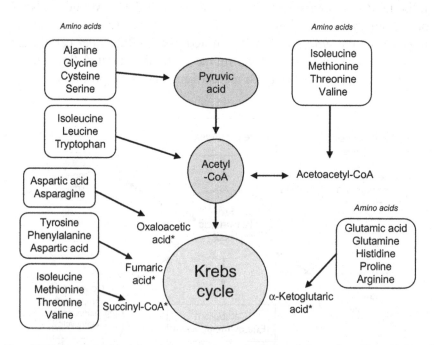

Figure 2.5 Amino acids (carbon skeletons) that can be converted to Krebs cycle intermediates (denoted by *) and used for energy metabolism.

reversible. Once it occurs, the amine nitrogen is formed into an ammonium ion (NH_4^+) and subsequently enters the urea cycle, where the product, urea, is excreted in urine.[2,4,5]

BCAA and Glucose Cycles

In addition to the anaplerosis means of providing intermediates for the Krebs cycle, there are several special considerations for some amino acids in ATP formation. The essential amino acids leucine, isoleucine, and valine, collectively termed the branched chained amino acids (BCAAs) because of the chemical structures, can be taken up directly and metabolized for energy in skeletal muscle. This use of the BCAAs for ATP production occurs during exercise.[2,4]

There are also the glucose—alanine and glucose—glutamine cycles. The amino acids alanine and glutamine are readily produced in skeletal muscle as part of the transamination reaction. They are released from the muscle during contraction (exercise), taken up by the liver from the blood, and used in the gluconeogenesis pathway, which results in the generation of glucose from noncarbohydrate carbon substrates. This glucose can be used as an energy source for ATP or stored as glycogen.

BIOENERGETICS

The amount of ATP generated through the various biochemical pathways varies according to (1) whether the pathway is anaerobic or aerobic and (2) which macronutrient is being used as an energy source (fuel). Fig. 2.6 illustrates this point.

Intuitively, it can be difficult to conceptualize molecular or even molar amounts of ATP. So the use of the kcal is a more popular means of thinking about how much energy is expended when a macronutrient is metabolized for energy. Table 2.1 presents representative values for the caloric (kcal) content of 1 g of each macronutrient.

The information in Table 2.1 reinforces what is shown in Fig. 2.6 at the cellular level; that is, lipid metabolically results in a much greater overall energy yield. Yet lipid cannot be relied upon entirely as a fuel source for exercise energy production; there must be some carbohydrate as well.[2,6] Chapter 3, "Regulation of Energy Metabolism During Exercise" introduces the regulatory aspects of controlling energy metabolism utilization and elaborates on this need for mixed fuel consumption.

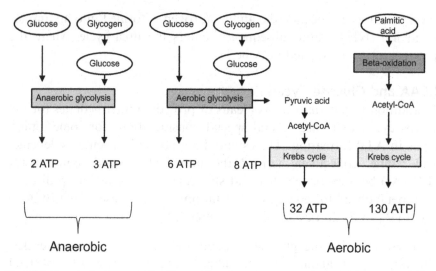

Figure 2.6 Approximate ATP production bioenergetics using various energy substrates and energy metabolism pathways.

Table 2.1 The Energy Content (kcal/g) of Macronutrients Used as Energy Sources			
Nutrient General	Nutrient Specific	Energy Content Chemical (kcal/g)	Energy Content Physiological (kcal/g)
Carbohydrate	Mixed	4.1	4.0
	Glycogen	4.2	
	Glucose	3.7	
Lipid	Mixed	9.3	9.0
	Palmitic acid	9.3	
Protein	Mixed	5.7	4.0
	Alanine	4.4	
	Isoleucine	6.9	
Chemical content is determined with a bomb calorimeter. Physiological content is in vivo in humans.[6–8]			

Close Up: What Am I Burning as an Exercise Fuel? How Chemical Analysis Can Determine What Is Being Used for Energy Metabolism

A classic method of assessing energy expenditure and what substrate (fuel) is being used as a source of the chemical energy for ATP production is indirect calorimetry. It involves analysis of exhaled respiratory gases by open-circuit spirometry (see photograph)[9] to calculate the respiratory quotient (RQ), which is the ratio of the volume of carbon dioxide

(CO_2) produced to oxygen (O_2) consumed. This analysis allows calculation of the amounts of carbohydrate and lipid being metabolized. The procedure is more than a century old and is still used extensively because it is inexpensive, accurate, and noninvasive.[10] It has been adapted over the years for use with various forms of exercise but is basically unchanged since the 19th century (see Fig. 2.7; photo is from around 1900).

Newer techniques to examine some of the same chemical elements noted above uses stable isotope analysis. For example, gas chromatography linked to isotope ratio mass spectrometry (GC/IRMS) is becoming common for gas analysis procedures. Liquid chromatography linked to IRMS (LC/IRMS) is a more recent biochemical technological development that broadens the range of compounds that can be targeted, in particular enabling the analysis of ^{13}C in nonvolatile aqueous soluble organic compounds such as carbohydrates and amino acids. The newer techniques provide greater accuracy and can assess individual compounds with a great deal of sensitivity. The trade-off in comparison to indirect calorimetry is much greater expense, which sometimes dictates whether a researcher can use them or not.

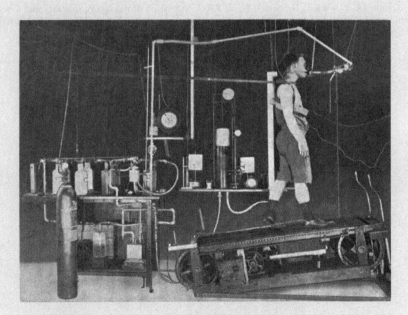

Figure 2.7 Indirect calorimetry system for measuring energy expenditure used during exercise testing nearly 100 years ago.

REFERENCES

1. *Health, United States, 2014 with special feature on adults aged 55–64.* US Department of Health and Human Services. Center for Disease Control and Prevention. Hyattsville, MD: National Center for Health Statistics; 2015.

2. Nelson DL, Cox MM. *Lehninger Principles of Biochemistry.* 6th ed. New York: Macmillan Higher Education; 2013.

3. Mitchell P. Coupling of phosphorylation to electron and hydrogen transfer by a chemiosmotic type of mechanism. *Nature.* 1961;191(4784):144–148.

4. Lemon PWR. Beyond the zone: protein needs of active individuals. *J Am Coll Nutr.* 2000;19 (Suppl 5):513S–521S.

5. Dunathan HC. Mechanism and stereochemistry of transamination. *Vitam Horm.* 1970;28:399–414.

6. Coyle EF. Fuel and fluid and fuel intake during exercise. *J Sports Sci.* 2004;22:39–55.

7. Jeukendrup AE, Saris WHM, Brouns F, Halliday D, Wagenmakers AJM. Effects of carbohydrate (CHO) and fat supplementation on CHO metabolism during prolonged exercise. *Metabolism.* 1996;45:915–921.

8. Benedict FG, Cathcard EP. *Muscular Work: A Metabolic Study with Special Reference to the Efficiency of the Human Body as a Machine.* Washington, D.C: Publication 187, Carnegie Institute of Washington; 1913.

9. Atwater WO, Rosa EB. A respiration calorimeter and experiments on the conservation of energy in the human body. *Annual Report Storrs Agriculture Experimental Station (Storrs, CT).* 1897;10:212–242.

10. Romijn JA, Coyle EF, Hibbert J, Wolfe RR. Comparison of indirect calorimetry and a new breath 13C/12C ratio method during strenuous exercise. *Am J Physiol.* 1992;263(1): E64–E71.

CHAPTER 3

Regulation of Energy Metabolism During Exercise

The aim of this chapter is to explain the mechanisms by which the body precisely and accurately regulates the energy metabolism pathways during exercise.

OVERVIEW OF CONTROL SYSTEMS

The physiological mechanisms that control energy metabolism are classified as either central or peripheral in origin. Central mechanisms are those components associated with the central nervous system (CNS) and are linked to the endocrine system, thus the term neuroendocrine system control is used. Peripheral mechanisms are those components at the level of the cells and tissues. This organizational dichotomy is not completely independent, as there is some degree of overlap and interaction in mechanisms as means of influencing energy metabolism during exercise. There are certainly elements of behavior and environment that influence exercise energy metabolism, but this chapter will focus on the physiological mechanisms.

Fig. 3.1 gives an overview of how the neuroendocrine system functions in energy metabolism. It shows that the principle mechanism to bring about change is enzymatic. Physiologically most of these enzymes exist in one of two forms, active and inactive. Regulation requires the changing of the activation status of enzymes or the formation of new enzymes by protein synthesis. In most situations, the active form is necessary to catalyze biochemical reactions to proceed and in this case lead to increased energy production. Activated enzymes can be associated directly with an energy producing pathway (eg, enzymes within glycolysis) or with a pathway that provides energy substrate (eg, enzymes within glycogenolysis → glucose) that an energy producing pathway can use.[1,2] Either of these means can lead to increased ATP production.

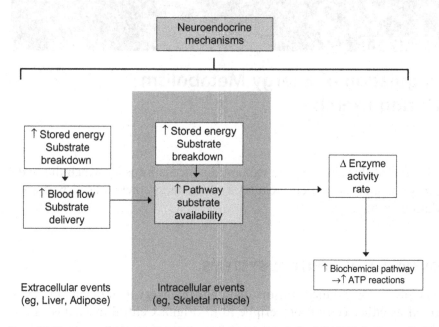

Figure 3.1 How neuroendocrine regulatory influences induce physiological and biochemical changes allowing increased energy production. Δ, *change.*

CENTRAL MECHANISMS—NEUROENDOCRINE

The Autonomic Nervous System (ANS)

The autonomic nervous system (ANS) is the fundamental regulator of energy metabolism. It consists of the sympathetic and parasympathetic subdivisions within the peripheral nervous system. The ANS influences the rate of ATP production in several ways (Table 3.1). The hypothalamus area of the brain is the overall integration center for the ANS—the "boss." The frontal lobes of the cerebral cortex also play a key role and act via the limbic lobe to influence hypothalamic function. Other controls come from the reticular formation in the brain stem and from the spinal cord. As the name implies, the ANS is an automatic system working at a subconscious level to ensure homeostasis (see below) and appropriate physiological adjustment to perceived or encountered stresses. The ANS, principally through the sympathetic subdivision, can change neuroendocrine input to tissues or organs that can invoke a multitude of physiological events affecting energy metabolism (Table 3.1).[3] The changes listed in the table are brought about by activation of adrenergic receptors on various tissues which respond to sympathetic stimuli.[4]

Table 3.1 Principal Influences of Increased Autonomic Nervous System Activity in Response to Homeostasis Disruption such as Exercise[6]

Physiological Process	Description of Physiological Process and Result
Vasoconstriction	Reduction in blood vessel diameter, reducing blood flow
Vasodilation	Increase in blood vessel diameter, increasing blood flow
Cardiac acceleration	Increase in the rate of heart contraction cycles
Myocardial contractility	Increase in the strength of heart muscular contraction
Bronchodilation	Increase in airway diameter of the lungs, increasing air flow
Calorigenesis	Increased production of heat via the digestion of food and/or by the action of certain hormones
Glycogenolysis	Increase in breakdown of stored carbohydrate into glucose for use in ATP production
Lipolysis	Increase in breakdown of triglycerides into glycerol and fatty acids for use in ATP production

Homeostasis is the maintenance of balance or equilibrium in the body's internal environment, even in the face of disruptive external or internal challenges. The body is constantly striving to maintain homeostasis.[5]

When there is a need for increased energy production, such as during exercise, sensory input to the brain increases sympathetic output (ie, afferent signals input, motor responses output) to specific tissues or organs. These sensory inputs can be conscious or subconscious and are a signal of changing homeostasis. The mechanisms of ANS sensorimotor events, detection-response, are extremely rapid and produce metabolic changes very quickly.[5]

The signaling agent of the sympathetic subdivision is the neurotransmitter norepinephrine (noradrenaline in the United Kingdom). The related hormone epinephrine (adrenaline) has many similar physiological effects. Hormones are the chemical messengers released into the blood, principally from endocrine glands located throughout the body. Norepinephrine and epinephrine are catecholamines (the major neurotransmitter dopamine is a catecholamine but is seldom discussed in the context of exercise). Epinephrine is secreted by the adrenal medulla endocrine gland, but its release is due to direct sympathetic innervation, and for this reason the gland is treated as an extension of the ANS. The adrenal medulla also secretes small amounts of norepinephrine into the bloodstream. The sympathetic nerve endings, which release norepinephrine, also contribute to blood levels as autonomic spillover occurs. That is, excess neural norepinephrine not taken up by target tissue diffuses into the bloodstream.[5,7]

Other endocrine glands also produce hormones that affect energy production at rest and during exercise. Table 3.2 provides a basic list and the functional effect of these major hormones associated with energy production and some of the specific tissues (targets) that these hormones influence. The magnitude and extent of the effect on the different processes influenced varies greatly from hormone to hormone.

Table 3.2 Energy Metabolism Roles of Selected Hormones of the Endocrine System on Tissues and Organs Essential to Exercise[8]		
Hormone (Source)	Target Tissue	Processes Influences
Glucagon (Pancreas alpha cells)	Muscle	↑ Glycogenolysis, Proteolysis (protein degraded)
	Adipose	–
	Liver	↑ Glycogenolysis, Gluconeogenesis, Proteolysis
Insulin (Pancreas beta cells)	Muscle	↑ Glycogenesis, Protein synthesis, ↓ Lipolysis
	Adipose	↑ Lipogenesis, ↓ Lipolysis
	Liver	↑ Glycogenesis
Epinephrine (Adrenal medulla)	Muscle	↑ Gluconeogenesis, Lipolysis
	Adipose	↑ Lipolysis
	Liver	↑ Glycogenolysis
Norepinephrine (Adrenal medulla + ANS)	Muscle	↑ Glycogenolysis, Lipolysis
	Adipose	↑ Lipolysis
	Liver	↑ Glycogenolysis
Growth hormone (Anterior pituitary)	Muscle	↑ Protein synthesis, Lipolysis
	Adipose	↑ Lipolysis
	Liver	↑ Gluconeogenesis
Cortisol (Adrenal cortex)	Muscle	↑ Lipolysis, Proteolysis
	Adipose	↑ Lipolysis
	Liver	↑ Gluconeogenesis
Testosterone (Testes ♂, Adrenal cortex ♀)	Muscle	↑ Protein synthesis
	Adipose	↑ Lipolysis
	Liver	↑ Protein synthesis, Glycogenesis
Estrogens (Ovaries ♀, Adrenal cortex ♂)	Muscle	↑ Lipolysis, Glycogenesis
	Adipose	↑ Lipolysis
	Liver	↑ Protein synthesis
Progesterone (Ovaries ♀, Adrenal cortex ♂)	Muscle	↑ Glycogenesis
	Adipose	↑ Lipolysis
	Liver	↑ Gluconeogenesis, Protein synthesis

Does the Brain Really Control Everything?

The simple answer is no. Organizationally, the CNS (consisting of the brain and spinal cord neural tissues) is viewed as a top-down system, but in actuality the regulatory influences at the local level of the tissue can be as critical or more so to induce changes in energy metabolism.[6] A key example is local blood flow to muscle during exercise. Metabolites (byproducts of metabolism) produced and released serve as humoral factors at the skeletal muscle and can dictate to a large degree how much blood flows through active musculature, potentially overriding a contradictory input from the CNS.[9] In this context it is important to remember that the blood can provide substrates for use in energy pathways and the oxygen necessary in the aerobic pathway, hence determining which pathways are more or less active.

PERIPHERAL MECHANISMS—INTERNAL MILIEU AND ALLOSTERIC MODULATION

Hormones and neurotransmitters are not the only regulators of enzyme activity controlling energy metabolism. A variety of molecules and ions produced within the cell are allosteric modulators in that they can activate or inactivate an enzyme and hence affect the overall rate of chemical reactions.[6,10] For example, during muscular contractions there are changes in ATP, ADP, magnesium (Mg^{++}), and calcium (Ca^{++}) levels which activate or inactivate key energy pathway enzymes. A change in the cellular levels of these substances can be immediate with the onset of muscle activity and is independent of, or acts in concert with, controlling signals from the central regulatory mechanisms.[6,10] Fig. 3.2 illustrates this process for the enzyme pyruvate dehydrogenase, a key regulator in the Krebs cycle within the mitochondrion.

The induction of enzyme activation by allosteric modulators speeds up the rate of ATP producing biochemical pathways and also the processes associated with producing more substrate for the energy pathways to use. For example, glycogenolysis can be accelerated, resulting in greater amounts of glucose for use in the glycolytic pathway.

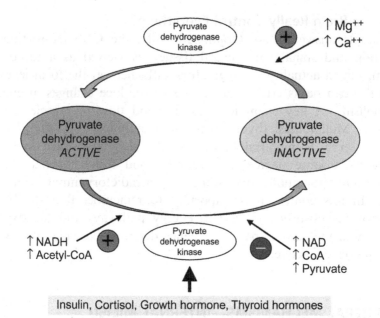

Figure 3.2 How allosteric modulation factors can influence the process of converting an inactive form of an enzyme into an active form (eg, the mitochondrial enzyme pyruvate dehydrogenase). +, stimulatory effect; −, inhibitory effect.

RUNNING OUT OF ENERGY

The central and peripheral mechanisms for controlling energy metabolism in healthy individuals are extremely sensitive and finely tuned. Nonetheless, many people experience a sensation of not having enough energy to do what they want when exercising. Regardless of the sensation, energy substrate is available and the metabolic pathways work. These people have not run out of energy. But the metabolic system can be challenged mentally to do more than it is currently capable off. This stresses the system and creates a homeostasis disruption that cannot be met at the current level of physical fitness (rate of ATP production desired > rate of ATP delivery capacity), and if continued can have negative consequences. So slowing down or stopping exercise is a fail-safe mechanism. As the body becomes more fit through exercise training, running out of energy becomes less severe and less frequent.

Close-Up: Tales from the USSR: The Transforming Research of Dr. Atko Viru of Estonia on Understanding the Regulation of Exercise Energy Production and Cellular Biochemical Adaptation

The Cold War began in 1947, shortly after World War II, and lasted until 1991. It was a state of political and military tension between the United States and its allies and the Soviet Union (USSR) and its allies. It was termed cold because there was no large-scale fighting directly between the two sides, although there were proxy wars in places such as Korea, Vietnam, and Afghanistan. Wars, proxy or otherwise, are fought on many fronts, and world sporting competitions were no exception. Starting with the 1952 Olympics Games in Helsinki, Finland and every 4 years afterwards, the United States and the USSR battled each other to determine whose athletes, ideology, and culture were the best.

The development of athletes and sporting teams is both an art and a science in which physiology, genetics, psychology, and intuition come together to help competitors reach peak performances at the most opportune times. Today, trainers and coaches working with US athletes have honed these skills to a near optimal level. But this was not always the case, especially in the application of science to sports. In the 1950s, 1960s, and into the 1970s, the Soviet sport scientists pushed the boundaries, and some of this work involved performance enhancing drugs that are now banned or illegal. But not all their science was focused on shortcuts. A dedicated group of Soviet scientists worked tirelessly to determine what training paradigms were critical, what periodization of training worked, and what level of cross training was necessary—things taken for granted now. One such scientist was Prof. Atko Viru (1933–2007), an Estonian who, when his country was part of the USSR (1945–1990), worked at Tartu University, the premier Soviet location to study sport science. Dr Viru was always an Estonian first, but during the Soviet occupation dedicated himself to determining how to improve performance in all athletes regardless of nationality. He published over 500 research papers on sport science and athletic training focusing primarily on the cellular biochemical adaptations that occur with training. In particular, he investigated the signaling agents that control many cellular events of metabolism and growth, namely hormones. His work was seminal in helping all in the sport sciences to understand why something was happening in response to training, which helped remove many aspects of the trial and error approach to developing athletes and improving performance.

I had the great good fortune of working with Dr Viru early in my career. He was a kind, gentle man of tremendous intellect. I found myself always challenged when talking to him, and needing to think carefully and critically about the work I was doing and what it meant. He always encouraged me to seek clear and concise answers. He was inspirational. Some years later, shortly before his death, I was having dinner with him and several leading sport scientists. Someone said that if there was ever a sports scientist who deserved the Nobel Prize, it was Atko Viru. Everyone voiced agreement. Dr Viru smiled, thanked us, and replied, "I am only as good as the students and colleagues I have been so fortunate to work with throughout my career." Many of us in the sport sciences owe a debt of gratitude to this humble man.

REFERENCES

1. Kahn BB, Alquiler T, Carling D, Hardie DG. AMP-activated protein kinase: ancient energy gauge provides clues to modern understanding of metabolism. *Cell Metab.* 2005;1(1):15–25.

2. Langfort J, Donsmark M, Ploug T, Holm C, Galbo H. Hormone-sensitive lipase in skeletal muscle: regulatory mechanisms. *Acta Physiol Scand.* 2003;178(4):397–403.

3. Kalsbeek A, Bruinstroop E, Yi CX, Klieverik LP, La Fleur SE, Fliers E. Hypothalamic control of energy metabolism via the autonomic nervous system. *Ann NY Acad Sci.* 2010;1212:114–129.

4. Brodde OE, Bruck H, Leineweber K. Cardiac adrenoceptors: physiological and pathophysiological relevance. *J Pharmacol Sci.* 2006;100(5):323–337.

5. Fuqua JS, Rogol AD. Neuroendocrine alterations in the exercising human: implications for energy homeostasis. *Metabolism.* 2013;62:911–921.

6. Hall JE. *Guyton and Hall Textbook of Medical Physiology.* 12th ed. Philadelphia, PA: Saunders; 2011.

7. Hasking GJ, Esler MD, Jennings GL, Burton D, Johns JA, Korner PI. Norepinephrine spillover to plasma in patients with congestive heart failure: evidence of increased overall and cardiorenal sympathetic nervous activity. *Circulation.* 1986;73(4):615–621.

8. Hackney AC, Lane AR. Exercise and the regulation of endocrine hormones. *Prog Mol Biol Transl Sci.* 2015;135:293–311.

9. Saltin B, Rådegran B, Koskolou MD, Roach RC. Skeletal muscle blood flow in humans and its regulation during exercise. *Acta Physiol Scand.* 1998;162(3):421–436.

10. May LT, Leach K, Sexton PM, Christopoulos A. Allosteric modulation of G protein-coupled receptors. *Ann Rev Pharmacol Toxicol.* 2007;47:1–51.

PART *II*

Applications: Knowledge into Practice

CHAPTER 4

Measurement Techniques for Energy Expenditure

The aim of this chapter is to present some of the widely used techniques for measuring energy expenditure during exercise in both laboratory and field settings, and to review their strengths and weaknesses.

HISTORICAL BACKGROUND

Measurement of energy expenditure is fundamental to the study of physiology and nutrition, and the proper choice and use of measurement methods is critical. The methodology can be traced back to Antoine Lavoisier's experiments on oxygen in the late 18th century. Modern assessment began with Wilbur Atwater and his colleagues Edward Rosa and Francis Benedict at Wesleyan University the 1890s. Technical aspects have advanced enormously, but the principles remain very close to those of Dr Atwater's time.[1]

Assessment can be done in one of two ways: directly with laboratory-based analytical procedures, and indirectly by estimation techniques in field situations.

LABORATORY BASED

Calorimetry is the science of determining the change in energy of a system by measuring the heat exchange of the system with the surroundings. For this discussion, the system is the human body. Results are typically reported in kilocalories (kcal), which can be expressed for total energy expenditure (TEE kcal) independent of time or expenditure relative to time (kcal/min or kcal/h).[2]

Direct Calorimetry

In human direct calorimetry a heat flow device measures the biologic heat released. Since the kcal is a thermal unit, the result can be readily converted to physiologic energy units. Direct calorimeters have to be large enough to contain a person, and, if exercise is involved, large enough to contain the exercise device and measurement equipment as well. There must also be a ventilation system to allow air flow in and out of the chamber for the duration of the assessment. These devices are expensive to construct and operate. Therefore most laboratories that do research on humans use the more cost effective indirect calorimetry.[3–5]

In nutrition and food sciences, the bomb calorimeter is a device used to assess the energy content of food items by direct calorimetry. The food item is combusted and the released heat quantified.[2]

Indirect Calorimetry

This is the method by which the rate of energy expenditure is estimated in vivo from total body respiratory gas exchange measurements—carbon dioxide (CO_2) production and oxygen (O_2) consumption—rather than directly from heat. The production of CO_2 and the utilization of O_2 are used in a mathematical formula to calculate oxygen uptake (VO_2).

$$VO_2 = [(V_E) \times (1 - (F_E O_2 + F_E CO_2)) \times (F_I O_2 / [1 - (F_I O_2 + F_I CO_2)]] - [(V_E) \times (F_E O_2)]$$

where:

V_E = volume of expired air ventilated per min;

$F_I O_2$ = fractional concentration of oxygen in air inspired;

$F_I CO_2$ = fractional concentration of carbon dioxide in air inspired;

$F_E O_2$ = fractional concentration of oxygen in air expired;

$F_E CO_2$ = fractional concentration of carbon dioxide in air expired.

This version of the formula is called the Haldane transformation.[6] The VO_2 being used is proportional to energy expenditure and is represented graphically in Fig. 4.1. An indirect calorimeter involves use of an open-circuit spirometry device to allow air (gas) volumes to be measured, and CO_2 and O_2 gas analyzers (see following sections for technical explanations) to determine the volumes of individual gases being produced or used.

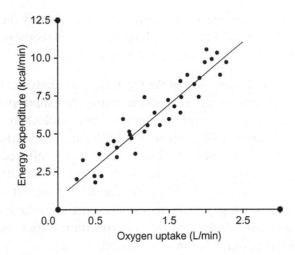

Figure 4.1 The change in an individual's energy expenditure (kcal/min) relative to increasing oxygen uptake (L/min).

Oxygen Uptake

The amount of O_2 consumed by the tissues is relative to the amount of ATP being produced by oxidative phosphorylation. "Oxygen consumption" refers to the O_2 being used at the cellular level, and "oxygen uptake" refers to that being used by the whole body as measured in respiratory gases. Both terms are abbreviated VO_2 and are used interchangeably, but they are technically different due to site of measurement. The O_2 uptake of a normal, healthy person of 70 kg body mass is ~250 mL O_2/min (3.5–4.0 mL O_2/kg/min expressed relative to body mass). VO_2 increases or decreases depending upon energy expenditure (Fig. 4.1).[4,6]

Gas Concentration Measurement

To determine VO_2 and hence energy expenditure, O_2 and CO_2 respiratory gas concentrations must be measured. The most common technique for O_2 in indirect calorimetry devices is the use of a paramagnetic cell, which works on the principle that O_2 is a strongly paramagnetic gas. That is, O_2 is attracted to a magnetic field because it has unpaired electrons in its outer electron ring. Most other respiratory gases are very weakly attracted. Most paramagnetic cell systems consist of a switched electromagnetic field and a pressure transducer.[6,7] The cell generates an electromagnetic field and gases are passed through the field. A pressure differential develops between the reference sample (usually room air) and the sample being analyzed.

The transducer detects the pressure fluctuations between the gases and converts the differential to a DC voltage which is directly proportional to the concentration of O_2 in the gas being analyzed.[7]

Several techniques are available for CO_2 gas analysis. In infrared absorption spectroscopy, molecules containing dissimilar atoms absorb infrared radiation and convert the energy into molecular vibrations. The vibration frequency depends on molecular mass and atomic bonding within the molecule. Most molecules will absorb infrared at specific wavelengths, hence the molecule can be identified and its concentration measured. Absorption is according to the Beer-Lambert law, which states that there is a logarithmic dependence between the transmission of light through a substance and the concentration of that substance.[7,8] This technique is perhaps the most common.

Refractometry also is used for CO_2. If monochromatic light is shown through a gas and focused on a screen, a pattern of light and dark bands appears at the fringe. The nature of these bands depends on the light waves arriving in or out of phase with each other, which in turn will depend on the gas refractive index and concentration. For example, in the Rayleigh refractometer, a series of prisms split the light source through tubes containing sample gas and control gas.[9] The refractometer is calibrated for a particular control gas at a known concentration. By aligning the fringe patterns created by each gas sample, a scale can be made to give the concentration of the unknown sample.[7–9]

Mass spectrometry is also used. A sample gas is drawn or injected into a low-pressure sampling chamber that is attached to another chamber at a pressure nearing that of a vacuum. The pressures are maintained by pressure–vacuum pumps. A molecular leak pathway exists between the two chambers. In the second chamber, the molecules that leak in are ionized, resulting in the loss of an electron. The resultant ions are accelerated by a cathode plate towards the second part of the chamber. At this part of the chamber either fixed magnets or electromagnets influence the ions and allow separation by the ionic mass and charge. The procedure can be used for other gases as well, such as O_2. In the quadrupole spectrometer, for example, the magnetic field is a mixture of a DC field and an AC radio frequency field.[10] If the AC component's frequency and the cathode acceleration are altered, only the ions of interest will be detected, as the others remain "trapped" in the magnet. These systems are highly accurate and require only extremely small amounts of gas samples.

Doubly Labeled Water (DLW) Technique Using Stable Isotopes

Doubly labeled water (DLW), developed in the early 1950s by Nathan Lifson and colleagues, is the gold standard to assess TEE due to its high degree of accuracy.[1,11] It can be used in a wide range of populations, even vulnerable ones such as pregnant and lactating women. It is highly suitable for use in a free-living context, is noninvasive, and imposes minimal burden to participants. Assessment is typically done over a period of several days, depending on the analytical approach necessary and the age of the participant.

DLW does not provide specific information on daily physical activity, though it does give an accurate measure of TEE over a chosen number of days or weeks, from which average daily energy expenditure can be calculated. It also does not quantify the activity type, intensity, or duration of energy expenditure. In use of the technique, daily urine samples are collected typically over 7–14 days and analyzed by isotope ratio mass spectrometry (IRMS). The stable isotopes deuterium (^2H) and oxygen-18 (^{18}O) are administered orally as a dose of drinking water, and the elimination of them from the body is tracked. The difference between the elimination rates of ^2H and ^{18}O is equivalent to the rate of CO_2 production, which can be converted to average TEE by standard stoichiometric equations.[8,10]

FIELD BASED

Calculation Formulas

The Harris–Benedict equation is used to estimate basal metabolic rate (BMR) and daily kcal energy requirements. The components used in the equation are anthropometric and age information. The estimated BMR value is multiplied by a number that corresponds to the individual's current or desired activity level. The resulting number is the recommended daily kcal energy intake to maintain current body mass, that is, weight (Table 4.1).

Table 4.2 presents alternative equations (Schofield) for calculating BMR. These equations have been used by the World Health Organization and the technical reports released by this agency.

Table 4.1 Harris–Benedict Equations for Determination of Basal Metabolic Rate (BMR) and the Typical Total Daily Energy Expenditure (TEE) with Varying Levels of Activity for Adults[12]

Harris–Benedict Equations
Men BMR = 88.362 + (13.397 × weight in kg) + (4.799 × height in cm) − (5.677 × age in years)
Women BMR = 447.593 + (9.247 × weight in kg) + (3.098 × height in cm) − (4.330 × age in years)

Multipliers for Increasing Physical Activity[a]	
Little or no exercise	Daily kilocalories needed = BMR × 1.2
Light exercise (1−3 days/week)	Daily kilocalories needed = BMR × 1.375
Moderate exercise (3−5 days/week)	Daily kilocalories needed = BMR × 1.55
Heavy exercise (6−7 days/week)	Daily kilocalories needed = BMR × 1.725
Very heavy exercise (2× per day, intensive workouts)	Daily kilocalories needed = BMR × 1.9

[a]*Underlined multiplier values are used to calculate TEE from calculated BMR.*

Table 4.2 Schofield Equations for Estimating BMR[13]

Age (years)	Males		Females	
	Equation (kcal/day)	Estimation Error (kcal/day)	Equation (kcal/day)	Estimation Error (kcal/day)
3−9	22.706 × M + 504.3	67	20.315 × M + 485.9	70
10−17	17.686 × M + 658.2	105	13.384 × M + 692.6	111
18−29	15.057 × M + 692.2	153	14.818 × M + 486.6	119
30−59	11.472 × M + 873.1	167	8.126 × M + 845.6	111
>60	11.711 × M + 587.7	164	9.082 × M + 658.5	108

M, *body mass in kg.*

There are many other such equations in the research literature, but a comprehensive listing is beyond the scope of this book.[4]

Physical Activity Questionnaires

In this methodology a person's reported level of daily activities is used to determine TEE based on behavioral patterns. The questionnaire is among the most widely used methods, and there are many types.[14−16] Global questionnaires are easy to administer and to complete, but they provide only minimal information about activity and simply enable a group to be categorized in general as active or inactive. Recall questionnaires are longer and provide more detailed accounts, including information about frequency and duration of activities over extended periods. The major differences between these questionnaires are the amount of detail, the length of time assessed, and the extent of supervision required for successful completion.[14–16] Duration of assessment varies from 24 h to 7 days or more.[14–16] These questionnaires are more valid when administered by interview, either on the telephone or face-to-face. Nonetheless, most such questionnaires are notorious for overestimating vigorous physical activity and underestimating time spent on activities of daily living.[16]

Activity Monitors
These devices are popular because they are convenient, relatively inexpensive, noninvasive, semiobjective, and versatile.

Heart Rate Monitors
Energy expenditure is estimated based on the assumption of a linear relationship between heart rate and VO_2. There is considerable interindividual variability in this relationship, but it is relatively consistent for an individual across a range of activities, and differences are predominantly a reflection of differences in movement efficiency, age, and physical fitness level.[4,15] The method, nevertheless, has limitations. For example, the relationship between heart rate and VO_2 differs between upper-body and lower-body muscular activities. And while there is a very close relationship between heart rate and energy expenditure during exercise, this is not the case during a state of rest or very light activity.[14,15]

Motion Sensors
Pedometers, though widely used, are limited in quantifying activity accurately. They work by registering "steps" taken during walking and running and give "counts" of the steps. Most of them fail to account for individual differences in physical make-up. Step count is also influenced by stride length (commonly related to height and leg length) and speed of walking.[15] For example, if an individual walks faster than normal, a pedometer may underestimate total distance. And it may overestimate distance for walking slower than is customary unless there are commensurate relative changes in stride length and step frequency when speed changes.

Accelerometers detect acceleration of the body. Acceleration is defined as the rate of change in velocity over a given time; therefore, the frequency, intensity, and duration of physical activity can be more readily assessed as a function of body movement with these sensors.[14,15]

The device consists of piezoelectric transmitters that are stressed by acceleration forces. This stress produces an electrical signal that is converted by processing units to produce an indication of movement. Those with microchips can convert the signal to kcal of energy expended based on mathematical formulae.[15] Accelerometry enables an estimation of intensity and duration of movement, and the relationship between accelerometer counts and energy cost allows physical activity to be classified by intensity.[15] The major advantages of an accelerometer are its relatively small size and the capacity to record data continuously over an extended period (days or weeks). Many research studies reported that they are objective, practical, noninvasive, accurate, and reliable tools to quantify physical activity volume and intensity with minimal discomfort.[15] Varieties include the simple uniaxial type that measures acceleration in one axial plane of motion, and the triaxial type with three planes of motion, which is far more precise.

LIMITATIONS AND ERRORS IN MEASUREMENTS AND ESTIMATIONS

Like all analytical procedures, laboratory and field-based techniques are subject to error. The sources are technical, analytical, and biological. Technical error is associated with the user and may be due to operational or calibration mistakes during setup. Analytical error is associated with the piece of equipment itself, such as its sensitivity and precision. Biological error is that encountered in the system (human) being measured, such as day to day variability in energy expenditure response (intraindividual) and different levels of response between individuals at rest and during various tasks (interindividual). It is important to recognize that error is additive, and the greater the total error, the less confidence can be had in the accuracy of the outcome measurement.[4,15,16]

Values reported for energy expenditure, laboratory or field based, represent the average expected response based on the input variables provided (eg, O_2 concentration, age, body mass). Too seldom is the margin of error reported or discussed in the literature with the specific techniques for measurement of energy expenditure. A simple rule of thumb is that laboratory-based techniques have significantly less error and greater accuracy than field-based techniques, provided that everything is done correctly.

Close-Up: Hot Parts of the Body: Use of Infrared Thermography in Exercise Energy Expenditure Research

According to the black body radiation law of physics, infrared radiation is emitted by all objects with a temperature above absolute zero. The amount of radiation increases as the temperature of the object increases. Infrared thermography registers the energy emitted, and this can be converted to temperature to develop a thermographic image showing the temperature distribution of an object such as the human body.

Credit for the concept of infrared radiation goes to the early 19th century British astronomer Sir William Herschel. He used a prism to deflect light from the sun and identified a range beyond the red part of the spectrum. This finding was confirmed and enhanced by placing a thermometer in the infrared range and noting the change in temperature. Herschel called these wavelengths calorific rays. They are now called infrared rays or infrared radiation.

Thermography has many commercial applications. Firefighters use it to see through smoke, to find people in fires, and to locate the base of a fire. Maintenance technicians use it to locate overheating joints and sections of power lines, which are at risk for failure. Construction technicians use it to find heating and cooling leaks in insulation and thus improve the efficiency of heating and air-conditioning units.

Another use of thermography is in exercise. Exercising muscle increases its surface temperature, which is proportional to the amount and intensity of the work being performed.[17] Sacripanti and associates developed an equation that uses radiant thermal energy to allow an estimation of the energy expenditure of an individual.[18] As an illustration of thermography, Fig. 4.2 shows images of an individual before, during, and after exercise. Deep red (neck and shoulders; for black-white images color is grey) coloration indicates the most intense infrared energy source, that is, the highest temperature.

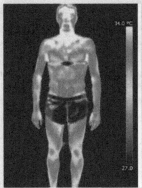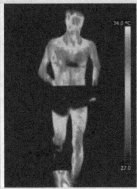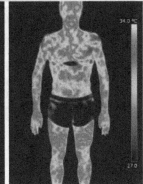

Figure 4.2 Thermogram of a person before (left), during (middle), and immediately after running 15 min (right). Used with permission.[17]

REFERENCES

1. Nichols BL. Atwater and USDA nutrition research and service: a prologue of the past century. *J Nutr*. 1994;124(Supplement 9):1718S–1727S.

2. Hargrove JL. History of the calorie in nutrition. *J Nutr*. 2006;136(12):2957–2961.

3. Webster JD, Welsh G, Pacy P, Garrow JS. Description of a human direct calorimeter, with a note on the energy cost of clerical work. *Br J Nutr*. 1986;55:1–6.

4. Hills AP, Mokhtar N, Byrne NM. Assessment of physical activity and energy expenditure: an overview of objective measures. *Fr Nutr*. 2014;1:5. Available from: < http://dx.doi.org/10.3389/fnut.2014.00005>.

5. Sahn DE, Lockwood R, Scrimshaw NS. *Techniques available for measuring energy expenditure. Methods for the Evaluation of the Impact of Food and Nutrition Programmes.* UN University; 1981: <http://archive.unu.edu/unupress/unupbooks/80473e/80473E0f.htm>.

6. Wilmore JH, Costill DL. Adequacy of the Haldane transformation in the computation of exercise VO_2 in man. *J Appl Physiol*. 1973;35(1):85–89.

7. Langton JA, Hutton A. Respiratory gas analysis: continuing education in anaesthesia. *Contin Educ Anaesth Crit Care Pain*. 2009;9(1):19–23.

8. IUPAC. In: McNaught AD, Wilkinson A, eds. *Compendium of Chemical Terminology, The "Gold Book"*. 2nd ed. Oxford: Blackwell Scientific Publications; 1997.

9. Allison J, Gregory R, Bich K, Crowder J. Determination of anaesthetic agent concentration by refractometry. *Br J Anaesth*. 1995;74:85–88.

10. Budzikiewicz H, Grigsby RD. Mass spectrometry and isotopes: a century of research and discussion. *Mass Spectrom Rev*. 2005;25:146–157.

11. Cole TJ, Coward WA. Precision and accuracy of doubly labeled water energy expenditure by two point and multipoint methods. *Am J Physiol*. 1992;263(5):E965–E973.

12. Roza AM, Shizgal HM. The Harris–Benedict equation reevaluated: resting energy requirements and the body cell mass. *Am J Clin Nutr*. 1984;40(1):168–182.

13. Schofield WN. Predicting basal metabolic rate, new standards and review of previous work. *Hum Nutr Clin Nutr*. 1985;39(suppl 1):5–41.

14. Shaopeng L, Gao R, Freedson P. Computational methods for estimating energy expenditure in human physical activities. *Med Sci Sports Exerc*. 2012;44(11):2138–2146.

15. Neilson HK, Robson RJ, Friedenreich CM, Csizmadi I. Estimating activity energy expenditure: how valid are physical activity questionnaires? *Am J Clin Nutr*. 2008;87:279–291.

16. Shephard RJ. Limits to the measurement of habitual physical activity by questionnaires. *Br J Sports Med*. 2003;37:197–206.

17. Tanda G. The use of infrared thermography to detect the skin temperature response to physical activity. *J Phys Conf Series*. 2015;655(1):012062.

18. Sacripanti A, Buglione A, De Blasis T, et al. Infrared thermography-colorimetric quantitation of energy expenditure in biomechanically different types of judo throwing techniques: a pilot study. *Ann Sports Med Res*. 2015;2(4):1026–1033.

Energy Expenditure at Rest and During Various Types of Physical Activity

This chapter focuses on those aspects of energy expenditure associated with daily living and the variety of physical activities that an individual encounters, specifically the components that are incorporated in total daily energy expenditure (TEE) and how those components are influenced to lead to more or less energy expenditure.

ENERGY EXPENDITURE AND UTILIZATION CONCEPTS

Earlier chapters introduced human energy needs, production, and use, and the basic concept that the more physical activity a person engages in, the greater is the energy expenditure. This chapter presents those aspects of energy expenditure associated with physical activities of daily living and of exercise. That is, the components of TEE and how they increase or decrease overall expenditure are addressed. The basic concept is nuanced by factors such as the source of energy (whether the adenosine triphosphate is from lipid or carbohydrate, whether new or body stores of these macronutrients are used as a fuel source), variability within a person from day to day, and variability between persons.[1,2] The terminology of the TEE concept is somewhat complex, but essentially the components consist of the "energy to exist" and the "energy of what we do."[2,3] Fig. 5.1 shows the components. To give an idea of baseline levels, Table 5.1 shows representative TEEs in sedentary people.

METABOLIC RATE IN THE RESTING STATE

A certain amount of energy must be expended just to exist, to keep all physiological metabolic processes working at a basal or resting level. This amount of energy is the basal metabolic rate (BMR), and measurement is typically taken under rigorously controlled conditions, usually by indirect calorimetry (see Chapter 4, "Measurement Techniques for Energy Expenditure"). It may occur in a darkened room upon

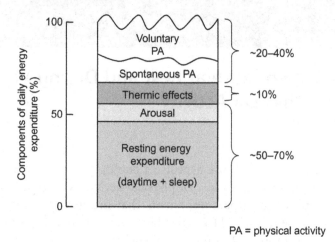

PA = physical activity

Figure 5.1 Components of TEE expressed as a percentage of daily energy contributions.

Table 5.1 Representative Daily Total Energy Expenditure (TEE) of Healthy Females and Males With Sedentary Lifestyle at Three Ages[4]

Age (years)	Gender	Weight (kg)	TEE (kcal/day)	TEE (kcal/min)
20	Female	60	1740	1.21
	Male	70	2160	1.50
40	Female	60	1620	1.13
	Male	70	1998	1.39
60	Female	60	1512	1.05
	Male	70	1836	1.28

waking after 7 to 8 h of sleep and 12 h of fasting (to ensure that the digestive system is inactive) with the subject resting in a recumbent position. BMR and resting metabolic rate (RMR) or resting energy expenditure (REE) are sometimes used interchangeably, but the latter two are taken under less restricted conditions, such as not requiring that the subject spends the previous night in the test facility.[2,5] For simplicity, RMR will be used in this book to refer to all three terms.

RMR can be influenced by many factors. Illness with fever and certain drugs can elevate values; advanced age and reduced caloric intake can reduce them. Body size and body composition are also important. For example, the value increases as body weight, height, and surface area increase. Adipose tissue (body fat) has lower metabolic activity than muscle tissue; therefore, increasing lean muscle mass increases values.[2,5]

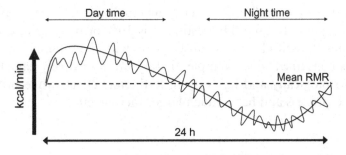

Figure 5.2 Changes in RMR over the course of 24 h.

The results of an RMR calculation is a number representing either the overall daily RMR (kcal/day) or the RMR rate (kcal/min). A healthy college-aged male might have a RMR of 1800 kcal/day (1.25 kcal/min).[4] This is not a constant value in a 24 h free-living period. Fig. 5.2 shows the average value as a dotted line. The actual value tends to oscillate minute by minute throughout a day (thin solid line in figure) and is influenced by wakefulness and arousal level.[5]

METABOLIC RATE DURING ORDINARY LIVING ACTIVITIES AND DURING EXERCISE

For the study of metabolism, the activity of ordinary free living is distinguished from physical activity or, the term adopted for this book, exercise. The more any of these activities are done and the more intensely they are done (ie, increased exertion), the great the amount of energy expended. Activity of free living is called spontaneous physical activity (SPA), defined as energy expenditure resulting primarily from unstructured mobility related to activities that occur during daily life. Exercise is referred to as voluntary physical activity (VPA), meaning it is more intentional in nature.[2,3,6]

THERMIC EFFECT OF FOOD

The thermic effect of food (TEF), also called specific dynamic action and dietary induced thermogenesis, is the amount of energy expenditure above the RMR due to the ingestion and digestion of food for use as energy or conversion to a storage form. It is one of the components of overall daily energy metabolism and energy expenditure. A commonly used TEF estimate is 10% of daily caloric intake, though the magnitude varies substantially with the different food constituents consumed. For example, dietary fat is relatively easy to process physiologically and has very little thermic effect, while protein is harder to process and has a much larger thermic effect.[2,5]

EXAMPLES OF ENERGY EXPENDITURE DURING VARIOUS ACTIVITIES

As the intensity at which an activity is done over time increases, the energy expenditure rate increases. Table 5.2 gives values for several different types of SPA and VPA of various levels of exertion.

Table 5.2 Energy Expenditure Rates for Exercises and for Activities of Daily Living (W = Watts, Unit of Power)[2]			
Activity	Energy Expenditure (kcal/min/kg)	Activity	Energy Expenditure (kcal/min/kg)
Exercises (VPA)			
Running (8 km/h)	0.12	Cycling (<16 km/h)	0.12
Running (9 km/h)	0.14	Cycling (16–19 km/h)	0.10
Running (10 km/h)	0.16	Cycling (20–23 km/h)	0.14
Running (11 km/h)	0.19	Cycling (24–26 km/h)	0.18
Running (12 km/h)	0.22	Cycling (27–31 km/h)	0.21
Running (14 km/h)	0.24	Cycling (stationary, 50 W)	0.05
Running (16 km/h)	0.28	Cycling (stationary, 100 W)	0.09
Walking (5 km/h)	0.06	Cycling (stationary, 150 W)	0.12
Walking (6 km/h)	0.07	Cycling (stationary, 200 W)	0.18
Walking (7 km/h)	0.08	Cycling (stationary, 250 W)	0.22
Circuit training	0.14	Calisthenics (push-ups, etc.)	0.08
Weight training (light)	0.05	Stretching/yoga	0.06
Weight training (hard)	0.10	Aerobics (low impact)	0.09
Rowing (50 W)	0.06	Aerobics (high impact)	0.12
Rowing (100 W)	0.12	Volleyball (recreational)	0.05
Rowing (150 W)	0.15	Swimming (light)	0.10
Rowing (200 W)	0.21	Swimming (moderate)	0.14

(Continued)

Table 5.2 (Continued)			
Activity	Energy Expenditure (kcal/min/kg)	Activity	Energy Expenditure (kcal/min/kg)
Activities of daily living (SPA)			
Chopping wood	0.09	Bathing/dressing	0.04
Mowing lawn (walking)	0.08	Sexual activity (kissing)	0.02
Raking leaves	0.07	Sexual activity (active)	0.03
Trimming (manual)	0.07	Ironing	0.04
Weeding/gardening	0.07	Making bed	0.04
Sitting activities (light)	0.03	Sleeping	0.02
Standing (light)	0.04	Child care (sitting)	0.05
Sweeping	0.05	Child care (standing)	0.06
Washing car	0.07	Carrying groceries (light)	0.07
House cleaning	0.06	Laundry folding	0.04
Washing dishes	0.04	Playing with kids (sitting)	0.04
Cooking food	0.04	Playing with kids (standing)	0.05

ESTIMATION OF EXERCISE ENERGY EXPENDITURE: METABOLIC CALCULATIONS AND FORMULAS

Many formulas to calculate energy expenditure are available on the Internet and require only the insertion of values (age, gender, estimated intensity, etc.) to obtain results. Some are well grounded in scientific research, others less so. The American College of Sports Medicine (ACSM), a respected scientific organization, has published the following for several common forms of exercise.[7]

Walking

$$\text{VO}_2 \text{ (mL/kg/min)} = (0.1 \times S) + (1.8 \times S \times G) + 3.5 \text{ mL/kg/min}$$

Running

$$\text{VO}_2 \text{ (mL/kg/min)} = (0.2 \times S) + (0.9 \times S \times G) + 3.5 \text{ mL/kg/min}$$

Stationary cycling (lower body—legs)

$$\text{VO}_2 \text{ (mL/kg/min)} = 1.8(\text{WR})/(\text{BM}) + 3.5 \text{ mL/kg/min} + 3.5 \text{ mL/kg/min}$$

Cycling (upper body—arm cranking)

$$\text{VO}_2 \text{ (mL/kg/min)} = 3.0(\text{WR})/(\text{BM}) + 3.5 \text{ mL/kg/min}$$

Step climbing

$$VO_2 \text{ (mL/kg/min)} = (0.2 \times f) + (1.33 \times 1.8 \times H \times f) + 3.5 \text{ mL/kg/min}$$

where:

> VO_2, oxygen uptake (mL/kg/min);
> S, speed (m/min);
> BM, body mass (kg);
> G, percent grade (hill) expressed as a fraction;
> WR, work rate or power (kg/m/min);
> f, stepping frequency/min;
> H, step height (m).

VO$_2$ can be converted to energy expenditure by multiplying by body mass and using a constant of 1000 mL of VO$_2$ being consumed equal to approximately 5 kcal/min.[2] A sample calculation:

Walking speed = 5 km/h → 5000 M ÷ 60 min = 83.3 M/min
Treadmill grade = 2%
Body mass = 70 kg
Duration = 30 min

$$\begin{aligned} VO_2 \text{ (mL/kg/min)} &= (0.1 \times S) + (1.8 \times S \times G) + 3.5 \text{ mL/kg/min} \\ &= (0.1 \times 83.3) + (1.8 \times 83.3 \times 0.02) + 3.5 \\ &= 14.8 \text{ mL/kg/min} \end{aligned}$$

$$\begin{aligned} \text{Energy expended (kcal/min)} &= (14.8 \text{ mL/kg/min}) \times (70 \text{ kg}) \\ &\quad \times (5 \text{ kcal}/1000 \text{ mL } VO_2) \\ &= 5.2 \text{ kcal/min} \end{aligned}$$

Total energy expended for activity = 30 min × 5.2 kcal/min = 156 kcal

FACTORS INFLUENCING RMR ENERGY EXPENDITURE

RMR can be influenced by a variety of factors. Some are major and persistent, such as aging. Others are minor and transient, such as a mild fever. Here are the principle ones and their effect:[2,8]

Age

RMR decreases with age, resulting in lower TEE unless there is a compensatory increase in SPA and/or VPA. The largest physiological factor with age is changing body composition. By the fourth decade of life, there is a gradual loss of muscle mass (sarcopenia) and function. Physically inactive people can lose 3% to 5% of muscle mass per decade.[2,9,10] Exercise reduces the rate of loss but does not eliminate it. Some causative factors of sarcopenia:

- A decrease in the number of spinal cord motor neurons and functioning motor units responsible for the innervation of skeletal muscle.[10]
- A decrease in the levels of some hormones such as growth hormone, testosterone, and insulin-like growth factor that are involved in muscle anabolism.[9,11]
- A decrease in the body's ability to synthesize protein, which is associated with an increase in myostatin, a myokine (ie, a cytokine or proteoglycan peptide-protein substance) produced and released by muscle cells (myocytes) that acts on muscle cells' autocrine function to inhibit myogenesis (muscle cell growth and differentiation).[9]
- Inadequate intake of calories and/or protein to sustain muscle mass.
- An increase in muscular apoptosis, the process of programmed cell death, via biochemical alteration, which leads to characteristic cell morphology changes and ultimately cell death (between 50 and 70 billion cells die each day by apoptosis in the average adult).[12]

Gender
Women's metabolic rate is approximately 5% to 10% lower than men's even when they are of the same body weight and height.[13] This slightly lower RMR is attributed to differences in androgenic-anabolic hormone status, skeletal muscle metabolism, and sympathetic nervous system activity. Body surface area may also play a role, as taller, thinner people usually have a somewhat higher RMR, and on average men are taller than women. There is some evidence that gender differences become less prominent as age increases.[12–14]

Obesity
Obesity is a condition of excessive body fat. The World Health Organization currently considers it to be at a pandemic level. It is associated with a state of chronic low-grade inflammation which increases the risk for many comorbidities.[15]

A comorbidity is the result of, or is strongly related to (ie, a risk factor for), a primary disease. Obese people are at increased risk for insulin resistance and type 2 diabetes mellitus, hypertension, dyslipidemia, cardiovascular disease, stroke, sleep apnea, gallblad-der disease, hyperuricemia and gout, and osteoarthritis.

In obesity the excess fat can be of the subcutaneous type (under the skin; visible) or the visceral type (around the organs; less visible). Metabolic energy expended by adipose tissue (fat cells) is lower than that of other tissues such as skeletal muscle. Hence, with increased body fatness adding to overall body mass, there is an increase in RMR, but to a much smaller extent than if muscle were being added.

Miscellaneous

A decrease in food intake can lower the metabolic rate as the body tries to conserve energy reserves. A very low calorie diet of fewer than 800 kcal/day reduces the RMR by more than 10%.[1,3] RMR can also be affected by drugs such as antidepressants, which can produce weight gain, and stimulants, which can produce weight loss. Drugs used to treat hypothyroidism increase RMR to a more normal level and restore a euthyroid state. Other aspects of the influence of pharmaceuticals and nutraceuticals on energy expenditure are discussed in chapter 8, "Pharmacologic and Nutritional Substances to Enhance Performance or Produce Weight Loss." Stress, fever-inducing illness, and some dia-betic states can elevate RMR.[16] Menopause may affect metabolism is such a way as to reduce RMR and lead to postmenopausal weight gain.[1–3,12]

Close-Up: Energy Expenditure Expression Confusion: What Exactly Is a Metabolic Equivalent Task (MET) and Why Should I Care?

Several units of measurement are used in energy expenditure—for exam-ple, kilocalorie (kcal), calorie, joule (J), kilojoule (kJ)—and this multiplic-ity can be confusing to the nonscientist. One that stands out in adding to the confusion is the MET, sometimes called simply the metabolic equiva-lent. It is a physiological measure expressing the energy cost of physical activities and is defined as the ratio of metabolic rate (therefore the rate of energy expended) during a specific physical activity to a reference RMR, set by convention to ~ 3.5 mL O_2/kg/min for adults:[17]

$$1 \text{ MET} = 1 \text{ kcal/kg}/h = 4.184 \text{ kj/kg}/h$$

The MET was developed several decades ago by a group of physical educators and physical activity specialist. The intent was to devise a

Table 5.3 MET Requirement for Common Activities at Three Levels of Intensity	
Activity	**MET Required**
Light intensity	<3
Sleeping	0.9
Watching television	1.0
Writing, desk work	1.8
Moderate intensity	3–6
Walking (4.8 km/h)	3.3
Light, easy calisthenics	3.5
Leisurely bicycling (<16 km/h)	4.0
Vigorous intensity	>6
Heavy, intense calisthenics	8.0
Rope jumping	10.0
Running (13 km/h)	13.0

means of expression for energy expenditure during activities that would be multiples of a person's resting energy status. For example, a two-MET activity required twice as much energy as being in a resting state. It was thought that this multiples-of-rest idea would be intuitively easier to understand and conceptualize. The intended target groups were nonscientists and scientists who were unfamiliar with measuring and expressing human energy expenditure. The unit was adopted by several leading health-related organizations, the America Health Association for one. It has remained a steadfast means of expressing energy expenditure of activities within the fields of nutrition and exercise epidemiology, though its popularity has waxed and waned over the years.

The conceptual value of the MET and similar energy metrics lies in their reinforcement of the fact that expressions of energy expenditure are interrelated and convertible from one unit to another with the use of simple mathematics. That is, the intent is to make some confusing scientific mathematics easier to comprehend by the general public. Table 5.3 gives examples.[2,17,18]

REFERENCES

1. Miller DS. Factors affecting energy expenditure. *Proc Nutr Soc*. 1982;41(2):193–202.

2. McArdle W, Katch F, Katch V. *Exercise Physiology: Energy, Nutrition, and Human Performance*. 5th ed. Philadelphia, PA: Lippincott Williams & Wilkins; 2001.

3. Garland T, Schutz H, Chappell MA, et al. The biological control of voluntary exercise, spontaneous physical activity and daily energy expenditure in relation to obesity: human and rodent perspectives. *J Exp Biol*. 2011;214:206–229.

4. Schofield WN. Predicting basal metabolic rate, new standards and review of previous work. *Hum Nutr Clin Nutr.* 1985;39(suppl 1):5–41.

5. Durnin JVGA. *Basal metabolic rate in man. Report to Food & Agriculture Organization/ WHO/United Nations University.* Rome: FAO Publishing; 1981.

6. Gerrior S, Juan W, Peter B. An easy approach to calculating estimated energy requirements. *Prevent Chronic Disease.* 2006;3(4):A129.

7. Glass S, Dwyer GB. *ACSM's Metabolic Calculations Handbook.* New York: Lippincott Williams & Wilkins; 2006.

8. Warwick PM. Predicting food energy requirements from estimates of energy expenditure. *Aust J Nutr Diet.* 1989;46(Suppl):s3–s28.

9. Fuqua JS, Rogol AD. Neuroendocrine alterations in the exercising human: implications for energy homeostasis. *Metabolism.* 2013;62:911–921.

10 Roth SM, Ferrell RF, Hurley BF. Strength training for the prevention and treatment of sarcopenia. *J Nutr Health Aging.* 2000;4(3):143–155.

11. Hackney AC, Lane AR. Exercise and the regulation of endocrine hormones. *Progress Mol Biol Translat Sci.* 2015;135:293–311.

12. Hall JE. *Guyton and Hall Textbook of Medical Physiology.* 12th ed. Philadelphia, PA: Saunders; 2011.

13. Ferraro R, Lillioja S, Fontvieille AM, Rising R, Bogardus C, Ravussin E. Lower sedentary metabolic rate in women compared with men. *J Clin Invest.* 1992;90(3):780–784.

14. Lazzer S, Bedogni G, Lafortuna CL, et al. Relationship between basal metabolic rate, gender, age, and body composition in 8,780 white obese subjects. *Obesity.* 2010;18(1):71–78.

15. *Obesity: preventing and managing the global epidemic.* Report of a WHO Consultation Group (WHO Technical Report Series 894); 2004.

16. Ruggiero C, Ferrucci L. The endeavor of high maintenance homeostasis: resting metabolic rate and the legacy of longevity. *J Gerontol: Ser A Biol Sci Med Sci.* 2006;61(5):466–471.

17. Ainsworth BE, Haskell WL, Whitt MC, et al. Compendium of physical activities: an update of activity codes and MET intensities. *Med Sci Sport Exercise.* 2000;32(suppl):S498–S516.

18. World Health Organization. *What Is Moderate-Intensity and Vigorous-Intensity Physical Activity?* <http://www.who.int/dietphysicalactivity/physical_activity_intensity/en/ >; Accessed 01.25.2016.

CHAPTER 6

Energy Storage, Expenditure, and Utilization: Components and Influencing Factors

The objective of this chapter is to explain the energy balance equation (EBE) by describing each component of the equation and the factors that influence it.

ENERGY BALANCE EQUATION

The EBE is straightforward and easy to understand. It accounts for how body mass (weight) changes with the consumption of food [energy intake (EI)] and the expenditure of energy [energy output (EO)]. It gives an indication of a person's energy storage (ES) status, typically over a period of 24 h or longer. There are three basic assumptions. First, body mass can be conceived as energy reserves or stores to be used for adenosine triphosphate (ATP) production and energy expenditure. Second, all of the chemical energy in the food and drink consumed (EI) adds to these energy reserves. Third, all of the energy expended (EO) reduce the reserves. The components of the equation are depicted as follows:[1]

$$\text{Energy intake} \rightarrow \text{Energy storage} \rightarrow \text{Energy output}$$
$$\text{or}$$
$$\text{Energy intake} - \text{Energy output} = \text{Energy storage}$$

Although the equations are straightforward, each element has associated with it aspects that account for variance in the EBE and contribute to some inconsistency in outcomes when used. Here are some examples:

- Humans consume carbohydrates, proteins, and fats for EI. After ingestion, the net absorption of these macronutrients is variable and incomplete, with fecal losses accounting for 2% to 10% of gross EI.

Net absorption of macronutrients also varies among individuals and is dependent on the specific food eaten, method of preparation, intestinal factors, and their age.[2,3]

- ES reflects the net total of body mass in the forms of carbohydrates, proteins, and lipids (fats). Total body mass includes these factors as well as water, bone, connective tissue, nervous tissue, and other tissues. In most adults, the lipid reserves in adipose tissue are the largest source of stored energy. Carbohydrate and protein stores have considerable water associated with them, while adipose has essentially none. Interestingly, carbohydrate intake has an impact on renal sodium excretion, which results in changes in the water content of extracellular fluid. Therefore, changes in body mass are expected when the macronutrient composition of the diet is altered even when the energy content of the diet is held constant.

- The rate of whole body energy expenditure reflects the fuel metabolized for growth, body maintenance, physical activity, pregnancy, lactation, and other needs. All of these processes can be affected by age, health, climate, nutritional status, and gender.

Put simply, some components of the EBE are not constant; so the mathematics in the equation is not always constant.[4]

In general, the EBE indicates that the energy balance status of the body can be in one of three conditions: neutral, positive, or negative. These conditions are depicted in Fig. 6.1.

In the neutral state, EI and EO are equal, hence ES is unchanged.

$$EI = EO; \text{ therefore ES is unchanged} \rightarrow \text{neutral EBE}$$

In the positive state, EI exceeds EO and the person is adding to ES. The result is weight gain.

$$EI > EO; \text{ therefore ES is increasing} \rightarrow \text{positive EBE}$$

In the negative state, EO exceeds EI and the person is taking away from ES. The result is weight loss.

$$EI < EO; \text{ therefore ES is decreasing} \rightarrow \text{negative EBE}$$

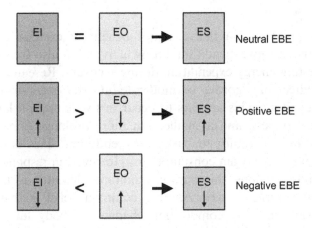

Figure 6.1 Schematic representation of the EBE, where EI is energy in (food consumed), ES is energy storage (body mass), and EO is energy out (energy expenditure) ($\downarrow$, decrease; $\uparrow$, increase).

EFFECT OF PHYSICAL ACTIVITY ON THE EBE

Effect on EI

Being physically active does affect EI. A misconception is that by being active one will automatically eat more; that is, an increase in EI will compensate for the developing negative energy balance. Research in which EO was increased by physical activity showed no significant compensatory change in EI—food intake did not increase—over 1 or 2 days.[1,3] Over a longer period, EI did gradually increase, but to a lesser degree than the increase in EO; that is, compensation was incomplete and a negative EBE developed.[1]

The EBE is used as a means of achieving weight loss in diet and exercise programs, but there is a great deal of individual variation in body weight (mass) response. Some individuals lose weight, others gain. Part of the variation may be due to a difference in exercise program adherence. But even when exercise sessions are closely supervised to eliminate this factor, variation remains.[5] Careful measurement of food intake before and after exercise suggests that part of the variation lies in how completely individuals do or do not compensate for their exercise prescriptions with higher food intake, which corresponds to the degree of initial hunger after an exercise session.[1]

Effect on EO

There is a popular idea that a major benefit of exercise comes not only from the energy expenditure during activity but also from an after-effect on resting energy expenditure during recovery. Research confirms a positive effect of vigorous or moderate-intensity exercise on resting energy expenditure. This appears to occur in two phases: a large effect that lasts 2 h or less, and a smaller but more prolonged one that can take up to 48 h to return to baseline expenditure.[6] This is called the excess postexercise oxygen consumption.[7] However, in response to most typical exercise sessions, this extra expenditure adds little to the overall EO occurring during a day. Another common belief is that exercise training results in body composition changes (↓ body fat, ↑ muscle mass) that generate an additional energy benefit of the exercise mediated through an increase in resting energy expenditure. In theory, this concept is sound, as muscle is more metabolically active than adipose tissue.[3] However, composition changes must be very large to have more than a minimal effect on EO. In most people this is not the case.

Exercise intervention programs may be counteracted by compensatory reductions in routine physical activity at other times of the day, although the information on this point is inconclusive. Some studies do report compensatory reduction in routine activities leading to no overall effect on daily EO; others report no compensation, and thus an increase in EO developed.[1]

These points emphasize an overarching generality, that all of the components of energy balance interact with each other. It is absolutely necessary to take all of these interactions into consideration when using the EBE, especially for calculating energy status with a goal to produce weight loss.

EXERCISE INTENSITY AND SUBSTRATE UTILIZATION

Fig. 6.2 shows that as exercise is done at greater intensity, the rate of energy expended (kcal/min) increases. It also shows the relative amounts of each fuel source used at each exercise intensity. As the intensity increases towards maximum, the contribution of carbohydrate as an energy source increases and the contribution of lipid decreases. This phenomenon of energy substrate shifting, the crossover concept, is explained in the next section.[8]

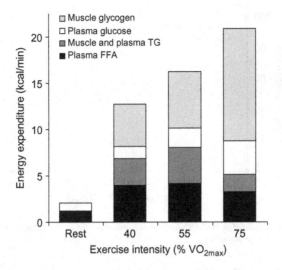

Figure 6.2 Energy expenditure rate at discrete increasing intensities of exercise [the proportional contributions of carbohydrates (glucose + glycogen) increase while those of lipids (TG = triglycerides + FFA = free fatty acids) decrease].[8]

THE CROSSOVER CONCEPT AND EXERCISE ENERGY METABOLISM

Fig. 6.2 shows the decrease in the use of lipid and the increase in the use of carbohydrate as exercise intensity increases. Fig. 6.3 shows this relationship across the continuum of exercise intensities. At rest, the contribution of lipid as a fuel is about 75%, and by maximal exercise it is <5%, while the pattern for carbohydrate is the opposite. The crossover point occurs at an exercise intensity of about 35% of maximal. Technically, the crossover point is defined as the muscular power output at which energy from carbohydrate-derived fuels predominates over energy from lipids, with further increases in power eliciting a relative increment in carbohydrate oxidation and a decrement in lipid oxidation.[9]

Why does this crossover occur? Increasing activity requires increasing contraction-induced muscle glycogenolysis, altered patterns of fiber type recruitment (greater numbers of fast-twitch vs slow-twitch), and increasing sympathetic nervous system activity. The oxygen delivery capacity of the cardiovascular system cannot keep pace with the large oxygen demand necessary to provide a continued reliance on lipid as an energy source.[9] These changes bring about more forceful and rapid muscular contraction and the generation of ATP more rapidly to allow the more intensive muscular activity to happen. Collectively these physiological

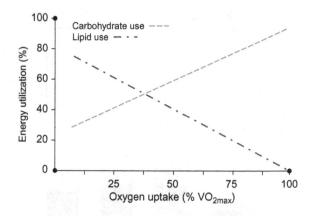

Figure 6.3 The crossover concept of how carbohydrate and lipid usage for energy changes as a person goes from rest to ever increasing exercise intensities.

adjustments dictate that the pattern of energy substrate use at any point in time depends on the interaction between exercise intensity induced responses (which increase carbohydrate utilization-oxidation) and exercise training-induced responses (which promote lipid utilization-oxidation; see Chapter 2, "Energy Metabolism of Macronutrients During Exercise"). As a side point, the terms utilization and oxidation are often used interchangeably, but they differ in biochemical meaning in energy metabolism. Utilization refers to the absolute amount of a substrate used [in grams (g)], while oxidation refers to the relative amount of substrate used (in percentages).

IMPLICATIONS OF THE CROSSOVER CONCEPT

There are important implications for exercise performance and weight loss. First, in people who engage in regular exercise, there is an improvement in the cardiovascular system's ability to deliver oxygen and muscle's ability to use oxygen.[10] This results in an improved capacity for aerobic energy metabolism. It also results in an increased reliance on lipid as an energy source during increasing intensity, and a sparing of carbohydrate, especially muscle glycogen. This point is critical, as there is an inverse relationship between muscle glycogen and the onset of fatigue with moderate to vigorous levels of exercise (Fig. 6.4).

Secondly, the illustration of the crossover concept in Fig. 6.3 suggests that lower-intensity exercise relies more, in percentage terms, on the breakdown of lipid as an energy source. This point is sometimes used to encourage people to engage in lower rather than higher-intensity exercise

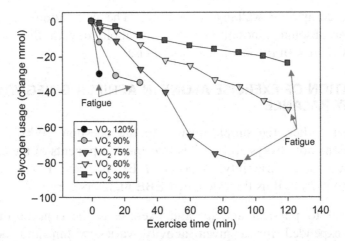

Figure 6.4 Muscle glycogen usage (mmol) at exercise of five intensities [intensity is expressed as a percentage of maximal oxygen uptake (VO₂ₘₐₓ)].[3]

Example 1	Example 2
• 45 min exercise session	• 45 min exercise session
• 30% VO_{2max} (walking)	• 60% VO_{2max} (running)
• VO_2 = 1.2 L/min steady state exercise	• VO_2 = 2.4 L/min steady state exercise
• VCO_2 = 1.042 L/min	• VCO_2 = 2.189 L/min
• RER = 0.868	• RER = 0.912
• Fat% = 55%, CHO% = 45%—RER Table[3]	• Fat% = 30%, CHO% = 70%—RER Table[3]
• kcal/L O_2 = 4.848—RER Table[3]	• kcal/L O_2 = 4.939—RER Table[3]
• 45 min x 4.848 kcal/L O_2 x 1.2 L/min O_2 = 261.8 kcal – Total energy expenditure	• 45 min x 4.939 kcal/L O_2 x 2.4 L/min O_2 = 533.4 kcal – Total energy expenditure
• 261.8 kcal x 0.45 CHO = 117.8 kcal	• 533.4 kcal x 0.70 CHO = 373.2 kcal
• 261.8 kcal x 0.55 fat = 144.0 kcal	• 533.4 kcal x 0.30 fat = 160.2 kcal
• 117.8 kcal CHO ÷ 4.02 kcal/g	• 373.2 kcal CHO ÷ 4.02 kcal/g
• CHO = 29.3 g utilized—Nutritional Table[3]	• CHO = 92.8 g utilized—Nutritional Table[3]
• 144.0 kcal fat ÷ 8.98 kcal/g	• 160.2 kcal fat ÷ 8.98 kcal/g
• Fat = 16.0 g utilized—Nutritional Table[3]	• Fat = 17.8 g utilized—Nutritional Table[3]

Figure 6.5 Energy expenditure from carbohydrate and lipid during a 45 min exercise session at low versus moderate intensity [L, liters, VCO₂, volume of carbon dioxide produced (used in RER calculation), RER, respiratory exchange ratio, CHO, carbohydrate].[3]

activities to lose body fat. However, evidence suggests that the total caloric expenditure rather than the percentage reliance on lipid is a more important factor influencing body weight change.[3] Certainly low-intensity exercise can be used to create a negative energy balance, but the time devoted to exercise must be expanded to allow the same expenditure of calories that can be achieved with moderate or vigorous exercise. To illustrate the point, Fig. 6.5 provides a mathematical example comparing walking to running. The calculations show that a greater absolute amount of fat is metabolized with the moderate intensity than with the low-intensity exercise.

LIMITATION OF EXERCISE ALONE IN ACHIEVING NEGATIVE ENERGY BALANCE

As noted earlier, the simple answer to the question "Why do EBE calculations not always add up?" is that the components of the calculation have some variability. Aspects of EI, ES, and EO vary from person to person, thus the calculated EBE numbers vary.

Critical to physical activity and exercise is the concept that the energy expended for a given activity varies within and between individuals. So there is inherent variability in the metabolic energy expenditure response. Furthermore, a person adapts to exercise training (see Chapter 7, "Exercise Training and Metabolic Adaptation"). One adaptation is improved mechanical efficiency to perform the exercise, which results in a reduced rate of energy expenditure. For example, before commencement of a training program, a 70 kg male might expend 310 kcal to jog 5 km in 30 min. Upon completion of the program, he might expend 290 kcal in that same 5 km 30 min jog. The body becomes more efficient and does not use energy at as high a rate in doing the same exercise. The capacity to do more exercise and expend greater amounts of energy, however, does increase.

ENVIRONMENTAL INFLUENCES ON ENERGY EXPENDITURE: GEOGRAPHY AND CLIMATE MATTER

Climate has a strong potential to influence energy expenditure because of the body's need to maintain a relatively constant temperature. Humans are homeotherms, "warm blooded" animals that strive to have a constant body temperature, in contrast to poikilotherms, "cold blooded" animals whose temperature adjusts depending on the environment. As a human moves out of the biological comfort zone of 35 to 38°C for core body temperature, physiological adjustments take place.[3,11] A lightly clothed adult in an indoor environment at an air temperature below 25°C will increase energy expenditure to generate heat by shivering (cold-induced thermogenesis), and at an air temperature above 35°C will also increase energy expenditure to reduce heat by sweating.[11,12]

The critical climate factors are air temperature, wind speed, and the radiant temperature of the surrounding materials. All these factors can come together to make an environmental situation such that body core temperature decreases or increases and physiological events are implemented to allow maintenance of temperature homeostasis; for example, shivering to induce extra heat or sweating to facilitate cooling. The control of the physiological adjustments is via the hypothalamus (our thermostat) and the autonomic nervous system.[3,11] As discussed in Chapter 3, "Regulation of Energy Metabolism During Exercise," the latter is also directly involved with many aspects of regulating and influencing energy metabolism. Another critical component to dealing with climatic factors is behavioral adjustments. Sensory cues indicating that the body is becoming cold or hot invoke powerful behavioral responses to accommodate appropriately (eg, put clothing on or take it off).[3]

Close-Up: Fad Diets and Physical Activity Myths: Infomercials Are Not Always 100% Correct!

The infomercial started in the United States. Originally it was a staple of late night television, outside of prime time, but beginning about the year 2000 it became more popular with advertisers and began appearing at all times of the broadcast day. By 2014, annual revenue was estimated at more than US$200 billion. Its questionable reputation arose from hawking products and services of questionable validity and offering contradictory information. Many of the products in infomercials are in the health and fitness sector, especially for weight loss. With the popularity of fad diets, lose-weight-fast plans, and schemes to tone muscles without work, many myths have developed and been fostered by infomercials and the "fitness experts" working in them.[13] Here are a few.

Myth: Fad diets will help me lose weight and keep it off.

Fact: Such diets are too numerous to list here, and new ones emerge frequently. The US National Institutes of Health clearly states that fad diets are not the best way to lose weight and keep it off. These diets often promise quick weight loss if you strictly reduce what you eat or avoid some types of foods altogether. Some of them may help you lose weight at first, but they are hard to follow, and most people who try them quickly tire of them and regain any lost weight. Also, they may not provide all of the required macro- and micronutrients and be potentially unhealthy.

Myth: "Low-fat" or "fat-free" means low calories.

Fact: A serving of low-fat or fat-free food is usually lower in calories than an equal volume of the full-fat product, but some processed low-fat or fat-free foods have nearly as many calories as the full-fat version, or even more. Also they may contain added flour, salt, starch, or sugar to improve flavor and texture after the fat is removed, providing additional calories.

Myth: If I skip meals, I can lose weight.

Fact: Skipping meals may make you feel hungrier and eat more than you normally would at the next meal. In particular, studies show a link between skipping breakfast and obesity.

Myth: Eating meat is bad for health and makes losing weight harder.

Fact: Research does not support this statement. Lean meat in small amounts can be part of a healthy weight loss plan. Chicken, fish, pork, and red meat do contain some cholesterol and saturated fat which in excess can be unhealthy; but meats also contain healthy nutrients like protein, iron, and zinc.

Myth: Dairy products are fattening and unhealthy and should be avoided.

Fact: Fat-free and low-fat cheese, milk, and yogurt are just as healthy as whole-milk dairy products and are lower in fat and calories. Dairy products are a good source of protein to build muscle and calcium to strengthen bones. Also, most milk and some yogurts have vitamin D added to help the body use calcium.

Myth: Lifting weights is not a good way to lose weight because it will make me "bulk up."

Fact: Lifting weights or doing other weight-bearing exercises such as push-ups regularly can help you build stronger muscles and reduce the risk of injuries. Only intense strength training, along with a certain genetic predisposition, can build large muscles.

Myth: "Going vegetarian" will help me lose weight and be a healthier person.

Fact: Many vegetarians do eat fewer calories and less fat than nonvegetarians, which results in a lower caloric intake. Also research has shown that vegetarian-style eating patterns are associated with lower levels of obesity, lower blood pressure, and reduced risk of heart disease. But vegetarians are not immune to becoming over-weight—like others, they can make poor food choices that cause weight gain, such as eating large amounts of foods that are high in fat or calories.

Myth: Doing exercise to lose weight only counts if I can do it for long periods of time.

Fact: US health agencies recommend that you should exercise 150 to 300 min each week. Researchers advise spreading the sessions out over the week, and that aerobic sessions as short as 10 min are effective in increasing energy expenditure and improving health.

Myth: Fast foods are always an unhealthy choice. You should not eat them when dieting.

Fact: Nutritionist indicate that fast foods are generally unhealthy and may cause weight gain. But if you choose to eat them, you can select those that are nutrient rich, low in calories, and small in portion size.

REFERENCES

1. Hall KD, Heymsfield SB, Kemnitz JW, Klein S, Schoeller DA, Speakman JR. Energy balance and its components: implications for body weight regulation. *Am J Clin Nutr.* 2012;95(4):989–994.

2. Miller DS. Factors affecting energy expenditure. *Proc Nutr Soc.* 1982;41(2):193–202.

3. McArdle W, Katch F, Katch V. *Exercise Physiology: Energy, Nutrition, and Human Performance.* 5th ed. Philadelphia, PA: Lippincott Williams & Wilkins; 2001.

4. Johnstone AM, Murison SD, Duncan JS, Rance KA, Speakman JR. Factors influencing variation in basal metabolic rate include fat-free mass, fat mass, age, and circulating thryroxine but not sex, circulating leptin, or triiodothyronine. *Am J Clin Nutr.* 2005;82:941–948.

5. Donnelly JE, Hill JO, Jacobsen DJ, et al. Effects of a 16-month randomized controlled exercise trial on body weight and composition in young, overweight men and women: the Midwest Exercise Trial. *Arch Internal Med.* 2003;163:1343–1350.

6. Speakman JR, Selman C. Physical activity and resting metabolic rate. *Proc Nutr Soc.* 2003;62:621–634.

7. LaForgia J, Withers RT, Gore CJ. Effects of exercise intensity and duration on the excess post-exercise oxygen consumption. *J Sports Sci.* 2006;24:1247–1264.

8. Brooks GA, Mercier J. Balance of carbohydrate and lipid utilization during exercise: the "crossover" concept. *J Appl Physiol.* 1994;76(6):2253–2261.

9. Luc J, van Loon C. Use of intramuscular triacylglycerol as a substrate source during exercise in humans. *J Appl Physiol.* 2004;97(4):1170–1187.

10. Brooks GA. Bioenergetics of exercising humans. *Compar Physiol.* 2012;2(1):537–562.

11. Kenney WL, Craighead DH, Alexander LM. Heat waves, aging, and human cardiovascular health. *Med Sci Sports Exerc.* 2014;46(10):1891–1899.

12. Westerterp-Plantenga MS, van Marken Lichtenbelt WD, Strobbe H, Schrauwen P. Energy metabolism in humans at a lowered ambient temperature. *Eur J Clin Nutr.* 2002;56(4):288–296.

13. National Institute of Diabetes and Digestive and Kidney Diseases (NIDDK). Weight-loss and Nutrition Myths. <http://www.niddk.nih.gov/health-information/health-topics/weight-control/myths/Pages/weight-loss-and-nutrition-myths.aspx >; Accessed on 01.15.2016

CHAPTER 7

Exercise Training and Metabolic Adaptation

This chapter discusses the metabolic adaptations that occur with exercise training as well as the principles of structuring a basic exercise program. The cellular and energy substrate changes as adaptive responses to such training are presented, with the focus on distinguishing between aerobic and anaerobic exercise.

EXERCISE TRAINING PRINCIPLES

Developing and carrying out an exercise training program for the general public requires close adherence to a set of principles to assure optimal adaptation and to minimize the risk of injury or other negative health consequences. Researchers associated with the US Department of Health and Human Services Centers for Disease Control and Prevention and the American College of Sports Medicine advocate the following elements to be manipulated for aerobic and anaerobic training:[1–4]

- Frequency—the number of days/week devoted to exercise sessions,
- Intensity—the strenuousness of the exercise session,
- Time—the duration of an exercise session,
- Type (mode)—the activity involved in performing the exercise,
- Volume—the total dosage of exercise training exposure,
- Progression—how the exercise routine is modified as adaption occurs.

Table 7.1 gives specifics for these six elements in an aerobic program and Table 7.5 in an anaerobic (resistance) program with the intent of improving the health and physical fitness level of an individual. Exercise training for competitive athletes to improve sporting performance uses the same principles but is focused to develop a substantially higher level of physical fitness and neuromotor skill.[5]

Table 7.1 Components to be Manipulated in an Aerobic Exercise Training Program
Frequency
• ≥ 5 days/week of moderate-intensity exercise • ≥ 3 days/week of vigorous-intensity exercise • Combination of both moderate- and vigorous-intensity exercise 3−5 days/week
Intensity
• Moderate: 40−60% heart rate reserve (HRR) (see Table 7.3) • Vigorous: 60−90% HRR
Time
• 30−60 min/day—moderate intensity[a] • 20−60 min/day—vigorous intensity[a]
Type
• Rhythmic, aerobic exercise (see Table 7.2) • Uses large muscle groups • Requires little skill • Enjoyable • Involves the specificity principle[b]
Volume
• = frequency × intensity × time (duration) of exercise • 1000 kcal/week • ≥ 150 min/week moderate intensity • ≥ 75 min/week vigorous intensity
Progression

• Gradual progression following the overload principle (see Fig. 7.1)
• As you adapt and improve physical fitness over weeks
  − Increase time/duration of each exercise session first
  − Increase frequency/intensity of exercise sessions after further adaptation

[a]One continuous exercise session or ≥ 10 min bouts over the course of a day. [b]Specificity principle says the body's response and adaptation is highly related to the type, frequency, and duration of the exercise.

EXERCISE TRAINING ADAPTATION

Training brings about a variety of beneficial biological accommodations. These changes allow the whole organism to function optimally to improve physical fitness and performance. They lead to an increase in work capacity which is manifested in the ability to perform more exercise in a given period of time, and at a higher intensity for a greater period of time (ie, greater levels of energy expenditure; Fig. 7.1).[3,6] These improvements are driven by a combination of physiological and morphological adaptations in the skeletal muscle, cardiorespiratory, autonomic nervous, and endocrine systems. The following sections summarize the fundamental changes in response to training. Organizationally, these responses are dichotomized into those associated more with aerobic, cardiovascular-type activities and those associated more with anaerobic, strength-type activities.[6]

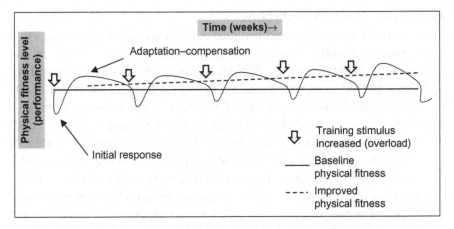

Figure 7.1 Progression of physical fitness levels in an exercise training program when the overload principle is employed.

Exercise Category	Exercise Description	Recommended Population	Examples
Table 7.2 Common Aerobic Exercises at Four Skill Levels[2,3]			
I	Aerobic (endurance) activities requiring minimal skills	Adults (sedentary)	Walking, leisurely cycling, aqua-aerobics, slow dancing
II	Aerobic (endurance) activities requiring minimal skills (potentially vigorous)	Adults (average or greater physical fitness level)	Jogging, running, aerobics, spinning, elliptical exercise, stair climbing (steps), fast dancing
III	Aerobic (endurance) activities requiring skills	Adults (average or greater physical fitness level)	Swimming, cross-country skiing, skating
IV	Recreational sport	Adults (average or greater physical fitness level and involved with regular exercise)	Racquet sports, basketball, soccer, netball, down-hill skiing, trail hiking

AEROBIC TRAINING

Table 7.2 presents common aerobic exercises at four skill levels, with specifics of each. The basic premise of aerobic exercise is that it strives to use large quantities of oxygen to produce adenosine triphosphate (ATP) by oxidative phosphorylation in skeletal muscle, hence a reliance on the cardiorespiratory system for oxygen delivery. The process incorporates dynamic movement of large muscle mass maintained at less than maximal effort and a relatively constant level of exertion (intensity). Hence the term steady state exercise session is commonly used to refer to the activity.

Cellular — Tissue Adaptations

The ability to use oxygen and produce ATP in the aerobic energy pathways depends greatly on the number, size, and efficiency of the mitochondria in skeletal muscle. Aerobic training increases their number and size, referred to as mitochondrial density. The increase in density is associated with increased activity of enzymes in oxidative pathways—the Krebs Cycle, β-oxidation, and the electron transport chain. Also, the enzymes and pathway components of glycolysis develop the ability to perform aerobic more than anaerobic glycolysis. An important corollary adaptation to increase mitochondrial density is improved capillary function for delivery of blood to the exercising muscles, hence providing more of what mitochondria need for oxidative ATP production, that is, oxygen. In the blood, hemoglobin concentration increases. Hemoglobin is found in the erythrocytes (red blood cells) and is the protein that carries 98% of the oxygen. These changes are part of the adaptations in the cardiovascular-respiratory (CVR) system. Another CVR change is a reduction in heart rate, the number of times the heart beats per minute. Following aerobic exercise training, the resting and submaximal exercise heart rates are lower, and the capacity of the heart to pump more blood per beat (stroke volume) is increased. This reduction of heart rate is brought about to a great extent by reduced sympathetic nervous system activity. Conversely, there is a compensatory increase in stroke volume, which allows a greater overall capacity of cardiac output (total volume of blood being pumped by the heart, expressed in L/min). Interestingly, reduced sympathetic activity lowers the degree of glycogenolysis occurring in skeletal muscle, which in turn permits greater lipid metabolism (see following section and Fig. 7.2).[6-8] Heart rate is a critical indicator of the level of work performed by the CVR, and is frequently used in aerobic exercise programs to monitor the intensity of the activity, since measurement of oxygen uptake (VO_2) is not practical. Table 7.3 presents a simple way for determining the level of exercise intensity using the heart rate reserve method (see Table 7.1, Intensity section, for recommendations of goals).[3]

Once oxygen is delivered, it moves from the blood into a muscle cell and binds to myoglobin, which is a protein very similar to hemoglobin but is found only in skeletal muscle and aids in shuttling oxygen to the mitochondria. Evidence is not conclusive but does point to aerobic training increasing the concentration of myoglobin, thereby allowing a great oxygen reserve to exist in muscle.[6,7]

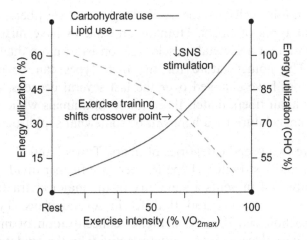

Figure 7.2 Exercise training induction of a rightward shift in the crossover point, resulting in greater reliance on lipid as an energy source (SNS, sympathetic nervous system; CHO, carbohydrate).

Table 7.3 Calculation by the Heart Rate Reserve Method for Determining Exercise Intensity

Heart Rate Reserve (HRR) Method—Example Calculation
40-year-old male
HR_{rest}: 60 beats per minute (bpm) HR_{max}: 180 bpm
Desired exercise intensity range: 60–70%
Formula: Target Heart Rate (THR) = $[(HR_{max} - HR_{rest}) \times \%$ intensity$] + HR_{rest}$
1. Calculate HRR $HRR = (HR_{max} - HR_{rest}) \rightarrow 180$ bpm $- 60$ bpm $= 120$ bpm 2. Convert desired exercise intensity into a decimal by dividing by 100 $60\% \rightarrow 0.60$ $70\% \rightarrow 0.70$ 3. Multiply intensity by HRR $0.60 \times 120 = 72$ bpm $0.70 \times 120 = 84$ bpm 4. Determine THR range by adding in HR_{rest} Lower limit: 72 bpm + 60 bpm = 132 bpm Upper limit: 84 bpm + 60 bpm = 144 bpm
Where: HR_{rest}, resting heart rate taken after a quiet period of rest (early morning preferred); HR_{max}, age-predicted maximal heart rate calculated from formula (220-age = HR_{max}); THR, target heart rate; HRR, heart rate reserve.

Skeletal muscle cells are the same thing as muscle fibers, and there are different types of fibers. Humans are said to have mixed muscle type, meaning that a skeletal muscle will consist of more than one type of fiber.[6] The nomenclature for the fiber types can be confusing, because names have changed over the last several decades, and some names of human fibers match those found in animals while others do not. The nomenclature used here will be somewhat simplistic.

There are two broad categories of fibers, Types I and II, also called slow-twitch and fast-twitch. Type II fibers are further divided into IIa and IIb. Table 7.4 presents a summary of the major characteristics of each fiber type. Types I and II differ in several ways. Type I has greater mitochondrial density, a higher concentration of myoglobin, greater capillary density, and a greater activity by the oxidative energy

Table 7.4 Comparison of Type I and Type II Muscle Fiber Characteristics			
Characteristics	Nomenclature		
	Type I (Slow-Twitch)	Type II (Fast-Twitch)	
	Type I: Slow Oxidative (SO)	Type IIa: Fast Oxidative Glycolytic (FOG)	Type IIb: Fast Glycolytic (FG)
Structural aspects			
Muscle fiber diameter	Small	Intermediate	Large
Mitochondrial density	High	Intermediate	Low
Capillary density	High	Intermediate	Low
Myoglobin content	High	Intermediate	Low
Functional aspects			
Twitch contraction time	Slow	Fast	Fast
Relaxation time	Slow	Fast	Fast
Force production	Low	Intermediate	High
Fatigability	Low	Intermediate	High
Metabolic aspects			
Creatine phosphate stores	Low	High	High
Glycogen stores	Low	High	High
Triglyceride stores	High	Intermediate	Low
Myosin-ATPase activity	Low	High	High
Glycolytic enzyme activity	Low	High	High
Oxidative enzymes activity	High	Intermediate	Low

producing enzymes. Consequently, Type I has a greater capacity for aerobic energy metabolism and resistance to fatigue as induced by continuous exercise activities (↑ endurance capacity). Type II, on the other hand, has lower aerobic capacity but is better suited for anaerobic energy production. There are slight differences between the IIa and IIb subtypes in function and capabilities, shown in Table 7.4.

There is controversy on whether a specific type or focus of exercise training completely changes one fiber type into the other. Evidence does firmly show that some fiber type characteristics are highly plastic and respond to exercise training dramatically, while others are more genetically determined and are essentially unresponsive to epigenetic stimuli such as exercise training. Interestingly, one of the more plastic responses is energy metabolism in relation to enzymatic function.[6,7]

Changes in Substrate Utilization

During many types and intensities of exercise, carbohydrate in the form of muscle and liver glycogen is the primary energy source (see Figs. 6.2, 6.3, and 6.4). Depletion of muscle glycogen is a major cause of fatigue, especially in prolonged aerobic activity (see later section this chapter and Fig. 6.4). Aerobic exercise training causes an adaptation in the utilization of energy fuels. For exercise performed at the same intensity, energy is produced by oxidizing less carbohydrate and more lipids (Fig. 7.2).[7-9] This modification is desirable because it allows the body to delay depletion of the limited stores of carbohydrate and hence the onset of fatigue. Every gram of lipid used essentially spares 2 g of carbohydrate because of the difference in the energy density of these metabolic substrates (see Table 2.1).[6,9]

Aerobic exercise training also improves the ability of the body to transport oxygen, that is, maximal oxygen uptake (VO_{2max}). As VO_{2max} is improved, a given submaximal workload represents a lower metabolic strain. For example, a running speed that elicited 60% VO_{2max} before training could be at 50% VO_{2max} afterwards. An increased VO_{2max} also allows a greater oxygen supply during submaximal exercise; because of increased mitochondria density, muscle can more efficiently use the oxygen (yielding greater aerobic ATP production). Collectively, these adaptations lead to greater ability to rely on lipid as a fuel source, as more oxygen is necessary to metabolize lipid than carbohydrate. The following illustrates this for the metabolism of 1 mol glucose (carbohydrate) versus 1 mol palmitic acid (lipid);

$$C_6H_{12}O_6 + 6\,O_2 \rightarrow 6\,CO_2 + 6\,H_2O + \sim 36\,ATP$$
$$C_{16}H_{32}O_2 + 23\,O_2 \rightarrow 16\,CO_2 + 16\,H_2O + \sim 130\,ATP$$

Lactate (lactic acid) is a byproduct of anaerobic glycolytic metabolism, and exercise at ever increasing intensities results in lactate accumulation in the blood. In response to exercise, everyone has a lactate threshold, the intensity of exercise at which the blood concentration of lactate begins to increase sharply (Fig. 7.3).[6,7] This point is also the maximal intensity of aerobic exercise that can be sustained for any long period of time. Above the threshold, muscular fatigue sets in, primarily through the effect of metabolic acidosis on muscle contraction and accelerated depletion of muscle glycogen. Aerobic training shifts the lactate threshold to higher intensities of exercise (Fig. 7.3) so that the body has a greater exercise capacity before fatigue occurs.[10] That is, one can run faster and further before having to slow down.[6,7]

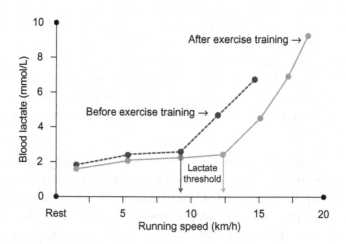

Figure 7.3 The exercise-induced rightward shift in the blood lactate threshold.

ANAEROBIC AND RESISTANCE TRAINING

Though aerobic and anaerobic activities have many similarities, they are distinctly different. Anaerobic is usually short-term (a few seconds or minutes) and engages large and small muscle mass at a higher intensity of activity than aerobic. In many situations, this type of exercise is not at steady state.

High-intensity interval training (HIIT) is now a popular form of anaerobic exercise. It might involve 30-s sprints at near maximum intensity followed by 30-s rests, or 1-min intensive runs followed by 2 to 3 min easy jogging, with a number of repetitive exercise bouts. Some HIIT designs are quite elaborate, but the basic principles and elements of the training were worked out by the German track and field coach Dr Woldemar Gerschler in the 1930s and have changed little.[11]

Resistance exercise—strength training—is not the same as HIIT. It involves skeletal muscle contraction exerting a force against an external load, as in weight lifting. Table 7.5 shows components to be manipulated in a resistance training program engaging small and large muscle groups, typically done by lifting weights with the intent of improving localized muscular endurance, strength, size (hypertrophy), or power.[1,3,12,13] These characteristics in Table 7.5 may be contrasted with the aerobic training characteristics in Table 7.1.

The basic premise of anaerobic or resistance exercise is that it involves the use of large quantities of stored ATP-PCr (ATP and phosphocreatine) and glycolysis as a means of ATP production (substrate phosphorylation) within skeletal muscle. These activities rely less on the CVR system for the delivery of oxygen and energy production unless the training principles are constructed to work more on the development of muscular endurance.

Table 7.6 explains the calculation of a useful metric, the 1-repetition maximum (1-RM), which is the maximum weight an individual is able to lift in a resistance exercise with good form through a full range of motion.[1,3,6,7,12,13] The 1-RM is used to prescribe the intensity of resistance based exercises, and is the corollary to the VO_{2max} in aerobic exercise.

Table 7.5 Components to be Manipulate in an Anaerobic Exercise (Resistance) Training Program

Frequency

- All major muscle groups 2–3 days/week
- Groups = chest, shoulders, upper and lower back, abdomen, hips, and legs
- 48 h separating training for same muscle group
- Split versus whole body workout—each is effective

Intensity

- 40–50% 1-repetition maximum (RM) sedentary individuals (see Table 7.6)
- 60–70% 1-RM novice to intermediate exerciser

Time

- No specific recommendation within a single exercise session

Type

- Multijoint exercises affecting >1 muscle group
- Target agonist and antagonist muscle groups (ie, opposed muscles)
- Single joint exercises targeting major muscle groups

Volume

- Sets: 2–4 per muscle group
- Rest: 2–3 min between sets[a]
- Repetitions: 8–12[a]

Progression

- Gradual progression of the frequency, intensity, or volume (repetitions and sets) of exercise (see Fig. 7.1)
- Progression approach may vary with goal of exercise training (see Table 7.7)

[a]*Repetitions (reps), the number of times to perform an exercise; sets, the number of times you will repeat the set number of repetitions.*

Table 7.6 A Safe Protocol for Determining the One-Repetition Maximum (1-RM) of a Resistance Exercise

Determining 1-Repetition Maximum (RM)

Maximum weight an individual is able to lift in a resistance exercise with good form through a full range of motion = 1-RM. A 1-RM test has an increased risk of injury since it assesses the maximum amount of weight a person can lift, which many people have never encountered; therefore, it is recommended that submaximal testing protocol be used to predict 1-RM. The following is one example.

Example protocol for bench press based on work of Brzycki:
Lie supine on flat exercise bench with feet pointing forward and back in neutral position. A spotter (helper person) should be used for safety.

1. Warm up: 5–10 repetitions at a low-resistance weight followed by 1 min rest.
2. Add 5–10 kg (or 5–10% of initial load); perform 10 repetitions.
3. Rest for 2 min between attempts before proceeding.
4. Repeat steps 2 and 3 until person is unable to perform 10 repetitions with proper form due to muscular fatigue.
5. Use the following equation:
   Predicted 1-RM = $1.0278 - 0.0278(X)$; where X is the number of repetitions performed in final step 4.

Source: *From* Brzycki M. Strength testing-predicting a one-rep max from reps-to-fatigue. *J Phys Educ Recreat Dance.* 1993;68:88–90.

Table 7.7 Illustration of the Principle of Specificity in Resistance Training

Specificity Principle: Resistance training response/adaptation is correlated to the type, frequency, and duration of exercise performed (recommendations for novices).
Question: What is the goal of the resistance exercise training program?

- Muscular endurance
  - Ability of a muscle or muscle group to repeatedly exert a submaximal resistance
  - Recommendation
    - Intensity: <70% 1-RM
    - Sets: 2–4
    - Repetitions: 10–25
    - Rest: 30 s–1 min
- Muscular strength
  - Ability of a muscle or muscle group to exert a maximal force
  - Recommendation
    - Intensity: 60–70% 1-RM
    - Sets: 1–3
    - Repetitions: 8–12
    - Rest: 2–3 min
- Muscular hypertrophy (increased size)
  - Enlargement of muscle size
  - Recommendation
    - Intensity: 70–85% 1-RM
    - Sets: 1–3
    - Repetitions: 8–12
    - Rest: 2–3 min
- Muscular power
  - Maximum power output attainable during a movement; explosiveness
  - Recommendation (highly dependent upon muscle group exercised)
    - Intensity: 0–60% 1-RM (0% intensity = exercises using body weight only)
    - Sets: 1–3
    - Repetitions: 3–6
    - Rest: 2–3 min

Table 7.7 illustrates the principle of specificity in resistance training, which is that adaptations are correlated with the type, frequency, and duration of the exercise, and shows recommended hypothetical program elements for progressively greater muscular development depending upon the goal of the program.[2,3,6] Since an individual's developmental goals for a resistance training program can vary greatly, examples in Table 7.7 are given for muscular endurance, strength, size (hypertrophy), and power development.

Cellular — Tissue Adaptations

Short duration (≤ 30 s) interval training improves the function of the ATP-PCr system for skeletal muscle energy development. The myosin-ATPase enzyme as well as those enzymes associated with creatine phosphate hydrolysis (eg, creatine kinase) is involved. This adaptation translates to enhanced muscular strength and power. Exercises requiring greater power output are highly affected by and dependent upon recruitment of Type II fibers, and the increased power of muscles after training is associated in part with improved neural recruitment patterns within the muscle.[6,7]

Longer duration interval training (eg, 1 to 4 min) at high intensities is still highly anaerobic, but there is greater aerobic contribution as exercise duration increases (see Fig. 1.2). These are typically not steady state exercise bouts. There is extensive interplay between use of anaerobic and aerobic glycolysis as a means of ATP production. Such training enhances activity of the key enzymes in the glycolytic pathway—glycogen phosphorylase, phosphofructokinase, and lactate dehydrogenase—leading to improved anaerobic–aerobic glycolytic ATP capacity. Longer duration interval training also is associated with major aerobic adaptations such as increased VO_{2max}, which is evidenced by improved mitochondrial function via increases in the key Krebs cycle enzymes such as succinate dehydrogenase and citrate synthase.[6,7,12,13]

Most resistance training programs are constructed to enhance strength and power, or to some degree enhance endurance. The former is associated with the muscle undergoing a greater degree of hypertrophy, that is, an increase in size, via increased muscle fiber cross-sectional area. A correctly constructed resistance program can increases fiber size within 2 months. The changes occur in part from increased synthesis of the contractile proteins (myofilaments) in the fiber. This enhanced anabolic state affects both Type I and Type II fibers, but Type II seem more responsive. Due to Type II characteristics (Table 7.4), this response leads to greater muscular strength and power. Interestingly, some evidence points to resistance training either having no effect on aspects of mitochondria (and capillaries) or actually decreasing the density and hence the aerobic capacity. By contrast, a program structured toward muscular endurance can improve aerobic capacity. That is, the specificity principle is still a dominant factor in adaptation; aerobic activities improve aerobic fitness more, anaerobic activities improve anaerobic fitness more (see the Close-Up below).[6,7]

Changes in Substrate Utilization
In short-interval training, the preferred energy substrate is stored ATP-PCr, but at longer intervals the substrate becomes carbohydrate in the forms of glycogen and glucose being used in the glycolytic pathway.

This can result in increased reliance on carbohydrate as the exercise energy source, and some evidence points to such training leading to increased muscle glycogen storage capacity. Yet if intervals are longer and the program is constructed to maximize aerobic development, then a shift towards increased reliance on lipid occurs. Whether this shift is to the same magnitude as seen in typical aerobic training programs (see above sections) is an issue of debate; the amount of research is limited.[6,7,9]

Several studies show that resistance training increases lipid utilization, especially with respect to postexercise (ie, the recovery period) energy substrate usage. However, the type and focus of the program—strength and power versus endurance—will greatly affect the magnitude of the substrate utilization adaptation. Importantly, the more aerobic-based the training, the greater the increase in lipid utilization, and the greater the potential to reduce body fat stores.[6,7,9]

MECHANISMS OF MUSCULAR FATIGUE

An age-old question: Why, when my brain tells my legs to move faster, do I slow down, or when I want one more repetition in a weight lifting set, I just cannot do it? The answer is muscle fatigue. Muscle fatigue can be defined as an exercise-induced reduction in the ability of the muscle to produce a desired force output or power.[6,14] A muscle can become fatigued for a variety of reasons related in part to the type of exercise (mode), intensity of the exercise, or the duration of the exercise. That is, slowing while sprinting 100 m at maximal effort and slowing while running a 42.2 km marathon involve different fatigue mechanisms. The followings are major mechanisms of muscular fatigue induction (see Hultman and Greenhaff for more extensive discussion).[14]

Reduced ATP availability in the skeletal muscle fiber reduces the ability of the contractile proteins actin and myosin (myofilaments) to interact, reducing sarcomere shortening and therefore force development. The sarcomere, the functional unit of a skeletal muscle fiber, is where tension develops at the cellular level and leads to muscular force output.[6] The fiber uses ATP to allow the cycling of the intracellular signal to initiate contraction, that is, calcium (Ca^{++}) moving in and out of its storage site the sarcoplasmic reticulum (SR).[6] If ATP production is not rapid enough or the amount of ATP is insufficient, the force of the fiber is reduced due to either compromised myofilament or SR action. This interplay of the rate and yield of ATP coming from the energy pathways can be a critical limiting factor in muscle function, especially for high power output contractions such as in sprinting 100 m.

Reduced PCr decreases the rapid rephosphorylation of adenosine diphosphate to ATP, a critical step in high power output activities. However, research results do not completely support reduced PCr as a strong factor in the development of fatigue. Reduced ATP and reduced PCr do result in accumulation of phosphate (Pi) in the fiber (see Chapter 2, "Energy Metabolism of Macronutrients During Exercise"), and the more extensive the muscular contractions the greater the Pi accumulation. Pi reduces the sensitivity of Ca^{++} responsiveness inside the fiber, resulting in a lessened intracellular signal for increased tension development and force production. Also, increased Pi inhibits force production by direct action on the contractile proteins (specifically at the myosin cross-bridges, critical for tension development). So any activity that produces a rapid accumulation of Pi could account for fatigue. High-intensity exercise readily does this.

Lactic acid accumulation has been proposed as a major cause of muscular fatigue for nearly 100 years. In muscle, lactic acid is produced in response to intensive exercise, and then it is rapidly broken down into lactate and hydrogen ion (H^+). If the lactate accumulation is rapid, there is the potential for intracellular pH levels to drop, and the resulting acidosis reduces enzymatic function. There is a strong temporal relationship between the development of fatigue and the accumulation of lactate and H^+, but the evidence does not support strong causality.[15] In fact, lactate production may be advantageous as it allows anaerobic glycolysis to proceed at a rapid rate [the production of lactate generates a critical enzymatic cofactor needed in an earlier glycolytic reaction (NAD^+; see Chapter 2, "Energy Metabolism of Macronutrients During Exercise")]. Furthermore, lactate can enter into the cell-to-cell lactate shuttle (the Cori cycle) and the intracellular lactate shuttle, where it can be used as an energy substrate.[7]

Muscle glycogen depletion can contribute to fatigue in prolonged exercise activities (see Chapter 6, Fig. 6.4) because of the reduced availability of substrate for ATP production ($\downarrow$ glycogen $\rightarrow$ $\downarrow$ glucose $\rightarrow$ $\downarrow$

glycolysis) and can lead to an impaired Ca^{++} release from the SR.[6,7] Persons participating in prolonged exercise at moderate to high intensity use a large amount of glycogen and have a high risk for development of severe fatigue late in the event ("hitting the wall" in marathon running, "bonking" in cycling). One of the bases for the use of sports drinks, which contain small amounts of carbohydrate, during exercise is to lessen this risk and allow performance to be maintained at a higher level. Carbohydrate loading, where the athlete increases dietary carbohydrate consumption to high levels for several days before an endurance event (ie, marathon) to increase muscle glycogen storage,[6] is done to prevent such occurrences as hitting the wall.

Prolonged or high intensity short duration exercise leads to the production of free radicals, atoms or molecules that contain one or more unpaired electrons that are capable of independent existence.[16] Free radicals are called reactive oxygen species (ROS). Skeletal muscles contain antioxidant substances to protect from excessive ROS production. Exercise can create oxidative stress, where the ROS formation exceeds the capacity of the endogenous antioxidants to maintain homeostasis. In such situations, the excess of ROS can disrupt charged particle exchanges across muscle cell membranes, enzymatic function, and contractile protein interactions, all leading to a cascading series of events inducing fatigue. This is a major reason why some exercisers take antioxidant dietary supplements (see Chapter 8, "Pharmacologic and Nutritional Substances to Enhance Performance or Produce Weight Loss").

In prolonged exercise, fatigue may also result from water loss or imbalance in the body. Dehydration of as little as 2% of body mass impairs performance. Loss in excess of 5% of body mass decreases the capacity for work by about 30%.[6] The main reasons that dehydration has an adverse effect are that it results in

- reduced blood volume,
- decreased skin blood flow,
- decreased sweat rate,
- decreased heat dissipation,
- increased core temperature,
- increased rate of muscle glycogen use,
- reduced maximal cardiac output.

The last point, reduced cardiac output, is a highly critical physiological consequence whereby dehydration impairs the CVR system and exercise capacity.[6,8,10] Also, dehydration is associated with impaired cognitive function and can affect judgment. Thus the maintenance of fluid intake during and after an exercise session is an important training aid and a safety step to prevent heat illnesses such as heat exhaustion and heat stroke.[3,6]

Close-Up: Cross-Training: Can You Get the Best of Both Worlds Biochemically?

Cross-training consists of combined exercise training with different exercise modes or sporting activities, usually by alternating regimens, as in running, weight lifting, bicycling, and swimming. Many people who cross-train will combine aerobic and resistance activities in the same session or on the same day. Aerobic and resistance exercises induce subcellular changes to bring about specific adaptations. Researchers have been exploring the cellular-molecular aspects of the signaling within muscle fibers to fully understand what activates specific adaptations.

Evidence suggests the presence of competing signaling cascades when aerobic and resistance training are done simultaneously, the concurrent training leading to an interference effect.

Heavy resistance training primarily sends a signal to a muscle fiber telling it to be highly anabolic, wherein the rate of protein synthesis exceeds the rate of protein breakdown, and over time the result is skeletal muscle hypertrophy. The fiber responds biochemically by phosphorylating the PI3K−Akt−mTOR signaling cascade and activating ribosomal protein s6 kinase (p70 s6k), both of which have been implicated in the anabolic molecular processes that follow acute and chronic resistance training. In contrast, aerobic training stresses a different signaling cascade which augments mitochondrial biogenesis, resulting in the formation of new mitochondria in the muscle fiber. The primary responses leading to this enhanced mitochondrial biogenesis are the activation of PGC-1α and AMP-activated protein kinase signal mechanisms.[17]

Research shows that, when done in isolation, both resistance and aerobic exercise accentuate these different cellular signaling mechanisms. But when resistance and aerobic training are done concurrently, there is suboptimal activation of both mechanistic pathways due to complex biochemical interplay in which activation and suppression of gene expression occurs and affects the other signaling mechanism. These findings suggest that cross-training does not maximize the full adaptive potential in muscle fibers for aerobic capacity or strength development—that is, the interference effect.

Several research groups are addressing this question, notably one led by Dr Keijo Häkkinen at Jyväskylä University in Finland, who is seeking to quantify and assess the interference as a function of the sequence of aerobic versus resistance activities in a single session and over multiple sessions (training). Work is ongoing, but preliminary results in sedentary people suggest that focusing on alternating days of aerobic-resistance exercise may be a way to get around the inference effect and better optimize adaptations.

REFERENCES

1. U.S. Department of Health and Human Services. *2008 Physical Activity Guidelines for Americans*. Atlanta, GA: USDHHS. <http://www.health.gov/paguidelines>; 2008.

2. Pate RR, Pratt M, Blair SN, et al. Physical activity and public health—a recommendation from the Centers for Disease Control and Prevention and the American College of Sports Medicine. *JAMA*. 1995;273:402−407.

3. American College of Sports Medicine. *ACSM's Guidelines for Exercise Testing and Prescription*. 9th ed. Philadelphia, PA: Lippincott Williams & Wilkins; 2014.

4. Garber CE, Blissmer B, Deschenes MR, et al. American College of Sports Medicine position stand: quantity and quality of exercise for developing and maintaining cardiorespiratory, musculoskeletal, and neuromotor fitness in apparently healthy adults—guidance for prescribing exercise. *Med Sci Sports Exerc*. 2011;43(7):1334−1359.

5. Bangsbo J. Performance in sports—with specific emphasis on the effect of intensified training. *Scand J Med Sci Sports*. 2015;25(Suppl 4):88−99.

6. Kenney WL, Wilmore J, Costill D. *Physiology of Sport and Exercise*. 6th ed. Champaign, IL: Human Kinetics; 2015.

7. Brooks GA. Bioenergetics of exercising humans. *Comparat Physiol*. 2012;2(1):537−562.

8. Clausen JP. Effect of physical training on cardiovascular adjustments to exercise in man. *Physiol Rev*. 1977;57(4):779−815.

9. Kiens B. Skeletal muscle lipid metabolism in exercise and insulin resistance. *Physiol Rev*. 2006;86(1):205−243.

10. Lundby C, Robach P. Performance enhancement: what are the physiological limits? *Physiology (Bethesda)*. 2015;30(4):282−292.

11. Gerschler W. Interval training. *Track Tech*. 1963;13:391−396.

12 Cadore EL, Pinto RS, Bottaro M, Izquierdo M. Strength and endurance training prescription in healthy and frail elderly. *Aging Disease*. 2014;5(3):183−195.

13. American College of Sports Medicine. American College of Sports Medicine Position Stand: exercise and physical activity for older adults. *Med Sci Sports Exerc*. 1998;30(6):992−1008.

14. Hultman E, Greenhaff PL. Skeletal muscle energy metabolism and fatigue during intense exercise in man. *Sci Progress*. 1991;75(298 Pt 3−4):361−370.

15. Posterino GS, Dutka TL, Lamb GD. L(+) lactate does not affect twitch and tetanic responses in mechanically skinned mammalian muscle fibers. *Pflügers Arch Eur J Physiol*. 2001; 442:197−203.

16. Nelson DL, Cox MM. *Lehninger Principles of Biochemistry*. 6th ed. New York: Macmillan Higher Education; 2013.

17. Hawley JA, Hargreaves M, Joyner MJ, Zierath JR. Integrative biology of exercise. *Cell*. 2014;159(4):738−749.

Pharmacologic and Nutritional Substances to Enhance Performance or Produce Weight Loss

This chapter discusses the effects of pharmacologic and nutritional supplements to enhance performance by altering biochemical processes or to produce weight loss by increasing energy expenditure. Relative to enhancing performance such items are sometimes called ergogenic aids.

The chemical agents used in this manner are many and varied and include ethical and decidedly unethical ones, traditional herbs, and clandestine synthetics. Performance-enhancing drugs (PEDs) are the subject of frequent sports scandals in the press. The whole topic is worthy of a complete book, and several excellent resources articles exist.[1,2] However, the field changes rapidly as new agents are brought to market, so that much information is outdated by the time it appears in print or online. Therefore, this chapter is a general guide and by no means exhaustive. For ease of comprehension, agents are presented in alphabetical order, with each explained according to its primary intent of use, either as a PED or a weight-loss agent.

ADAPTOGENIC HERBS

Adaptogens are nutraceuticals that supposedly stabilize physiological processes and promote tolerance to stress.[3,4] They are very popular and are often marketed with far-reaching health claims such as increased longevity, libido, and overall well-being. Many claimed to treat medical conditions, but none assessed thus far by the U.S. Food and Drug Administration (FDA) has proven effective and safe enough to gain approval in this capacity. Certain adaptogens in a few exercise studies have shown some positive results, such as increased work capacity, but overall the evidence is not very persuasive.[5]

Some prevalent examples are

- *Eleutherococcus senticosus* is a woody shrub in the Araliaceae family native to Northeast Asia. It is also called Siberian ginseng, eleuthero, and ciwujia. It is promoted as having a wide range of benefits, including boosting mental performance and making cancer-related chemotherapy more effective. It is used to treat chemotherapy induced bone marrow suppression, angina, hypercholesterolemia, neurasthenia, insomnia, and poor appetite.[3,4]

- *Lepidium meyenii* is a cruciferous plant native to the high Andes. It is also called Peruvian ginseng or maca. It is claimed to increase strength, energy, stamina, libido, and sexual function.[6]
- *Ocimum tenuiflorum* or *Ocimum sanctum* is in the Lamiaceae family and is native to the Indian subcontinent. It is also known as holy basil and tulasi or thulasi. It is supposed to enhance the body's natural response to deal with physical and emotional stress.
- *Panax ginseng* belongs to the Araliaceae family and is found in the cooler climates areas of North America and eastern Asia. It is used to improve mood, cognition, and immunity, and as an antifatigue agent. It is thought to provide nonspecific protection against mental, physical, and environmental stress.
- *Rhodiola rosea* is in the Crassulaceae family grows in cold regions throughout the world. It is also called golden root, rose root, roseroot, western roseroot, Aaron's rod, Arctic root, king's crown, lignum rhodium, and orpin rose. It is used for its purported antistress and fatigue-fighting properties and as an antioxidant.
- *Schisandra* is a twining shrub in the Schisandraceae family native to Asia and North America. It is also called magnolia vine. It is used to decrease fatigue, counter stress, and enhance physical performance. It is said to reduce stress hormones in the blood.
- *Withania somnifera* is in the Solanaceae family and is found in the drier regions of central Asia. It is also known as ashwagandha, Indian ginseng, poison gooseberry, and winter cherry. It is used for arthritis, anxiety, insomnia, tumors, tuberculosis, asthma, the skin condition leukoderma, bronchitis, backache, fibromyalgia, menstrual problems, hiccups, and chronic liver disease.

This is a small sample of herbal adaptogens and their uses. For greater detail, including proposed mechanisms of action, see selected references.[2-5]

AMINO ACIDS

Amino acids are the basic chemical constituents of protein, which is one of the three dietary macronutrients. Amino acids are common nutritional supplements used in exercise programs, especially by athletes doing resistance training. The premise for use is that skeletal muscle growth and adaptation depends upon adequate amounts of amino acids for building structural, enzymatic, and contractile proteins (ie, maintaining an adequate free amino acid pool). Theoretically, the principle is sound, in that inadequate dietary protein compromises the ability to adapt to the stimulus of exercise; and research evidence clearly shows the need for exercising people to consume more protein that sedentary ones.[7] But the more-is-better philosophy is inappropriate here and is faulty reasoning in this situation. When consumption of amino acids exceeds the ability of body tissues to incorporate them into protein structures, they remain in the free amino acid pool, where they are urinated off or converted to carbohydrate or lipid.[1] Hence just providing greater amounts of amino acids does not push the biochemistry to force greater muscle tissue protein synthesis. A physiological need and a regulatory stimulus are required (see the Close-Up of Chapter 7, "Exercise Training and Metabolic Adaptation").

ANABOLIC STEROIDS

Anabolic-androgenic steroids (AAS) are drugs biochemically and structurally related to testosterone, the major male sex hormone.[8] AAS serve to signal the process to increase protein synthesis within cells, especially in skeletal muscle (anabolic action). They also have androgenic and virilization properties, the development and maintenance of masculine characteristics.[1,8] These drugs were developed in the 1930s and are used therapeutically to stimulate muscle growth and appetite, induce male puberty, and treat chronic wasting conditions such as those secondary to cancer.[8] Research shows that they increase body weight, lean body mass, and muscular strength when used with exercise and proper diet.[1,8]

However, long-term use or high doses have major negative health effects, including changes in cholesterol level (increased low-density lipoprotein, decreased high-density lipoprotein), hypertension, liver damage, and cardiac abnormalities. And they produce conditions related to reproductive hormonal imbalance such as gynecomastia and reduction in testicular size. They and their derivatives have been prevalent in sports for decades.[1,8] Early research findings on their

Table 8.1 Major Side Effects Associated With Anabolic-Androgenic Steroid Use
Acne
Aggressiveness
Altered electrolyte balance
Alterations in clotting factors
Alterations in libido
Altered thyroid function
Clitoral enlargement (women)
Decreased endogenous testosterone production
Decreased gonadotropin production
Decreased high-density lipoprotein
Depressed immune function
Dizziness
Depressed spermatogenesis
Edema
Elevated blood glucose
Elevated blood pressure
Elevated creatine kinase and lactate dehydrogenase
Elevated total cholesterol
Elevated triglycerides
Gastrointestinal distress
Gynecomastia and breast tenderness
Increase apocrine sweat gland activity
Increased nervous tension
Liver toxicity
Lower voice (women)
Masculinization (women)
Muscle cramps/spasms
Nosebleeds
Polyuria
Premature closure of epiphyses (children)
Prostatic hypertrophy (and perhaps cancer)
Psychosis
Wilms tumor risk

effectiveness were equivocal, leading some scientists to claim that they did not work, while most athletes who used them claimed the opposite. Subsequent research has definitively shown that they are effective but are associated with many adverse side effects (Table 8.1).[8,9]

AAS are banned by all major sports governing bodies. Nonetheless they remain among the most common substances detected by the World Anti-Doping Agency (WADA).[10]

CAFFEINE

Caffeine is a central nervous system stimulant from the methylxanthine class. It is the world's most widely consumed legal psychoactive drug, that is, a chemical substance that changes brain function and results in alterations in perception, mood, or consciousness. It produces physiological effects by several mechanisms, the most prominent being reversible blockade of the adenosine receptor and consequent prevention of drowsiness induced by adenosine accumulation in the brain. It also stimulates aspects of the autonomic nervous system, in particular the sympathetic division, which affects catecholamine release and circulating levels of these hormones in the blood. This catecholamine-like effect can change energy metabolism and accelerate energy expenditure (see Chapter 3, "Regulation of Energy Metabolism During Exercise").[11–13]

Caffeine has both positive and negative health effects. In premature infants, it is used to treat bronchopulmonary dysplasia and prevent dyspnea. One meta-analysis (a form of research where large numbers of study findings are consolidated) concluded that coronary artery disease and stroke are less likely in people who drink 2 to 5 cups (1 cup = 237 mL) of coffee per day but more likely with over 5 cups consumption; a cup of brewed coffee typically contains 80 to 175 mg of caffeine, though the content can go much higher.[14] Some people experience insomnia or sleep disruption if they consume caffeine, especially in the evening, but others show little disturbance. Caffeine can produce a mild form of dependence, associated with withdrawal symptoms such as sleepiness, headache, and irritability. Tolerance to the stimulatory autonomic effects, such as increased blood pressure, heart rate, and urine output, develops with chronic use. The FDA classifies caffeine as generally recognized as safe. The lethal dose is 10 or more grams per day for an adult, roughly 50 to 100 cups of coffee. But pure powdered caffeine, available as a dietary supplement, can be lethal in very small amounts.[12]

EPHEDRINE

Ephedrine is a sympathomimetic amine and substituted amphetamine commonly used as a stimulant, mental concentration aid, decongestant, appetite suppressant, and treatment for the hypotension associated with surgical anesthesia.[15] It is similar in molecular structure to phenylpropanolamine and methamphetamine, and to the hormone epinephrine; hence it mimics many of their actions physiologically.[15,16] Ephedrine affects adrenergic receptors like the endogenous catecholamines (epinephrine and norepinephrine). Biochemically it is an alkaloid with a phenethylamine skeleton, which is found in various plants in the genus *Ephedra* (Ephedraceae family).[15] The herb *Ephedra sinica* (Chinese medicine name Ma Huang) contains ephedrine and pseudoephedrine as its principal active constituents.

Ephedrine promotes modest short-term weight loss, specifically fat loss, and is used by some body builders to reduce body fat before a competition.[13] The catecholamine-like action leads to increased performance in athletes, and ephedrine is synergistic with caffeine, thus the use of ephedra and caffeine together serve as a PED.[5,11] Ephedrine is potentially dangerous; by 2004 the FDA had received over 18,000 reports of adverse effects, and its sale in purified form was prohibited in the United States in 2005.[16]

ERYTHROPOIETIN—BLOOD DOPING

Prolonged aerobic exercise is highly dependent on the functionality of the cardiovascular system and a person's VO_{2max}.[1] Blood doping is an illicit method of improving performance by artificially boosting the blood's ability to bring more oxygen to muscles. The oxygen carrying capacity of the blood is directly related to amount of hemoglobin in red blood cells and the number of red blood cells. Blood doping is usually done in one of two ways: use of the synthetic recombinant human erythropoietin (rhEPO) or blood transfusion.[17] The process of making new red blood cells, erythropoiesis, is stimulated by the endogenous hormone erythropoietin (EPO). Taking synthetic EPO, originally developed to treat severe anemia, accelerates erythropoiesis and greatly improves blood oxygen carrying capacity and aerobic exercise performance.[17]

Transfusion was the first method of blood doping. Initially homologous transfusion was used, but because of its higher medical risk it was replaced by autologous transfusion. In the latter, the individual has quantities of their blood removed, stored, and reinfused weeks later after the erythropoiesis process has replaced the blood removed.[1,17] Either method improves blood oxygen carrying capacity.

Blood doping has serious health risks, the most critical being related to the increased viscosity of the blood adding to the work of the heart.

An estimated 20 competitive cyclists have died as a result of blood doping over the past 25 years.[17] Other risks are

- blood clot,
- heart attack,
- stroke.

Blood doping via transfusion carries additional risks such as

- HIV infection,
- hepatitis B,
- hepatitis C,
- bacterial infection.

The use of rhEPO also carries risks such as

- hyperkalemia (dangerously elevated blood potassium level),
- high blood pressure,
- mild flu-like symptoms.

LEVOTHYROXINE

The chiral compound levothyroxine (L-thyroxine) is a synthetic thyroid hormone chemically identical to thyroxine (T_4), which is endogenously secreted by the thyroid gland. It is used to treat thyroid hormone deficiency and occasionally to prevent the recurrence of thyroid cancer. Its enantiomer dextrothyroxine (D-thyroxine) was used in the past for hypercholesterolemia (elevated cholesterol levels) but was withdrawn from the market because of cardiac side effects.[18]

Thyroid hormones T_4 and T_3 (triiodothyronine) are highly effective in accelerating basal metabolic rate and hence energy expenditure (Fig. 8.1). They are also involved in the growth and development of tissues, including skeletal muscle. They are permissive, in that they facilitate the actions of other hormones such as epinephrine, glucagon, and growth hormone. Levothyroxine produces weight loss, enhances physical performance, and promotes more rapid recovery from exercise training. Interestingly, some athletes who do intensive training to develop pseudo-hypothyroidism, which can be mitigated by levothyroxine supplementation.[18,19] This course of action, however, must be done only under a physician's supervision.

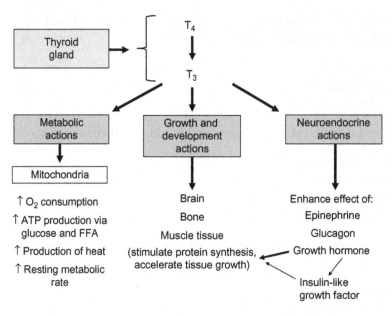

Figure 8.1 Physiological actions of thyroid hormones to affect metabolism, growth and development, and neuroendocrine events (T_4 = thyroxine, T_3 = triiodothyronine).

METRELEPTIN

Metreleptin is a synthetic analog of the adipocyte hormone leptin. It is used to treat diabetes and various forms of dyslipidemia.[20] Leptin is released from adipocytes and is called the satiety hormone because it helps to regulate energy balance by inhibiting hunger. These action of leptin is opposed by ghrelin (also called lenomorelin), the hunger hormone, which is produced primarily in the gastrointestinal tract. Both hormones act on receptors in the arcuate nucleus of the hypothalamus to regulate appetite and thereby maintain energy homeostasis.[18] In some forms of obesity there is decreased sensitivity to leptin (increased leptin resistance) and a resultant inability to detect satiety despite high energy stores, which leads to overeating. Leptin replacement drugs such as metreleptin were initially devised to treat hyperphagia (overeating) in leptin deficiency individuals.[21]

Leptin affects other physiological processes, as evidenced by its synthesis in cells other than adipocytes and the existence of leptin receptors on cells outside of the hypothalamus. These roles are not entirely clear and are under investigation.[21] Nonetheless, there are many instances illustrating the point where blood leptin levels are dissociated from the role of communicating energy reserves to the hypothalamus[22]:

- Leptin level is decreased after short-term fasting (<72 h), even when changes in body fat mass are not observed.
- In obese patients with obstructive sleep apnea, leptin level is increased; the increase is reduced by continuous positive airway pressure.
- Leptin level is reduced by sleep deprivation.
- Leptin level is increased by perceived emotional stress.
- Leptin level is decreased by elevated testosterone and increased by elevated estrogen.
- Leptin level is chronically reduced by exercise training.
- Leptin release is increased by the drug dexamethasone (a glucocorticoid).
- Leptin level is increased by insulin (↑ risk for type II diabetes).
- Leptin level is chronically increased in obesity.

SIBUTRAMINE

Sibutramine (brand names Reductil, Meridia, Siredia, Sibutrex) is an oral anorexiant. Anorexiants are thought to work by suppressing the appetite center in the hypothalamus, specifically by inhibiting the reuptake of serotonin and norepinephrine. Until 2010, sibutramine was prescribed as an adjunct for obesity, along with diet and exercise. It was withdrawn from the market in several countries because of an association with increased incidence of dangerous cardiovascular events, including stroke. In the United States, it is a Schedule IV controlled substance.[9] Drugs and certain chemicals used to make drugs in the United States are classified into five categories or schedules depending upon the drug's acceptable medical use and the its abuse or dependency potential. The abuse rate is a determinate factor in scheduling. Schedule I drugs are considered the most dangerous, with a high potential for abuse and severe psychological and/or physical dependence. Schedule V drugs have the least potential for abuse.[23]

VITAMINS AND MINERALS

Vitamins and minerals are dietary micronutrients and are necessary to transform the potential energy in food to chemical energy (adenosine triphosphate, ATP), but they are not direct sources of energy. Vitamins are essential organic compounds that function as regulators of protein, carbohydrate, and lipid metabolism to produce useable energy. Minerals are inorganic elements and act as cofactors for enzymes that influence all aspects of energy metabolism. Their roles as enzymatic regulators and cofactors influencing the rate of chemical reactions make them critical in the biochemical pathways of ATP production (see Chapter 2, "Energy Metabolism of Macronutrients During Exercise").[15]

There is no compelling evidence to support the general use of vitamin and mineral supplements to improve physical performance, although exercise training may increase the need for some of them.[1,11,24] In people eating a well-balanced diet with adequate caloric intake (neutral energy balance; see Chapter 6, "Energy Storage, Expenditure, and Utilization: Components and Influencing Factors"), the risk of inadequate vitamin and mineral intake is low. Vitamin and mineral supplementation may, however, improve the nutritional status of a person who consumes insufficient calories and nutrients (ie, someone in a chronic negative energy balance), thus enhancing exercise performance.[24]

For many years there has been a debate on the efficacy of supplemental vitamins C and E and β-carotene (a precursor of vitamin A) as antioxidants to enhance performance and promote health. Some evidence suggests an indirect beneficial effect of supplementation by reducing skeletal muscle damage and enhancing some aspects of immune function, again especially in individuals with improper nutritional habits.[1,24,25]

People who restrict their energy intake to participate in certain sports (eg, to qualify for a weight category in wrestling) run a greater risk of micronutrient deficiencies.[1,11] Girls and women participating in ballet, gymnastics, and endurance running are especially prone to inadequate intake of calcium and iron. Low-dietary calcium may increase the risk of stress fractures of the bones; hence supplemental calcium may promote optimal peak bone mass in these people. Inadequate iron may reduce performance in activities relying greatly on the cardiovascular-respiratory system, as hemoglobin has an iron component that binds and carries oxygen. Low body iron stores may be attributed to low iron intake or increased iron loss due to menstruation, gastrointestinal bleeding, and sweating, and hence may decrease aerobic performance. To this end, dietary supplementation can increase body iron stores and decreases lethargy in anemic people.[24]

The indiscriminate use of vitamin and mineral supplements in the absence of nutrient deficiency is unwise and can be detrimental. For example, chronic ingestion of large doses of vitamin C can induce renal stone formation, decreased blood coagulation time, produce red blood cell rupture (erythrocyte hemolysis), and lead to gastrointestinal disturbances. Excessive supplemental zinc can induce a secondary copper deficiency and decrease high density lipoprotein cholesterol (the "good" cholesterol). In short, for well-nourished people, long-term consumption of vitamin and mineral supplements in amounts exceeding the recommended daily allowance can result in adverse effects on some other nutrients, and supplement usage should be viewed judiciously.

WORLD ANTI-DOPING AGENCY TESTING PROGRAM

A brief history of the WADA formation is given in the Close-Up below. This important sports agency works to implement and enforce the World Anti-Doping Code to ensure a level playing field for all athletes. The stated purposes of the Code and the WADA program are

- "To protect the Athletes' fundamental right to participate in doping-free sport and thus promote health, fairness, and equality for Athletes worldwide; and
- To ensure harmonized, coordinated, and effective antidoping programs at the international and national level with regard to detection, deterrence, and prevention of doping."[10,26]

Detection focuses on ensuring that athletes are not using any of the substances on the WADA prohibited list. This list is extensive and regularly updated. The process involves secure urine and blood specimens being collected from an athlete at in- and out-of-competition settings by a WADA representative. Specimens are brought to a WADA-accredited laboratory, of which there were 32 as of 2014, for biochemical analysis. State-of-the-art bioanalytical techniques are used to assess samples. To maintain sample integrity, WADA has strict policies and procedures on the handling of samples to ensure there is no contamination or falsification. When sample analysis is complete, the results are forwarded to the sports association of the athlete. The athlete is then notified of the test results and of any sanctions if warranted by a positive finding of banned substances.

WADA itself does not impose sanctions. Other national and international sports agencies (eg, the United States Anti-Doping Agency and the International Olympic Committee) do that. Sanctions for positive doping tests range in severity from public warning to multiyear suspension from participating in competitions. Other common sanctions include "loss of results," which means that the athlete's results, record, or standing in competitions may be canceled or voided. In many cases, an athlete who won a medal or prize in a competition and subsequently tests positive must surrender the medal or prize and have the victory declared null and void. In serious cases, athletes may be barred for life from further sports participation.

Close-Up: Performance-Enhancing Drugs: Not Something New Under the Sun

The use of PEDs to enhance athletic performance is not a new phenomenon. The ancient Greeks used special diets and stimulating potions to fortify themselves in their Olympic games. Throughout the 19th and early 20th centuries, strychnine, caffeine, cocaine, and alcohol were often used, especially by runners and cyclists. In the 1904 St. Louis Olympic Games marathon, Thomas Hicks of the United States (gold medalist) took raw eggs, injections of strychnine, and doses of brandy throughout the race. This was viewed as perfectly acceptable by the authorities at that time.

In the 1920s sports governing bodies started to implement restrictions on the use of drugs and other agents by athletes because of health concerns. These restrictions, however, remained essentially ineffective, as no testing or screening was done to confirm or refute use. The development of synthetic hormones such as testosterone later in the century made matters worse, as these agents gained widespread use. Pressure on world sports authorities to introduce regular and rigorous testing was increased, in part due to the death from PED use of a Danish cyclist at the 1960 Rome Olympics.

In 1967 the International Olympic Committee (IOC) took action and founded its Medical Commission, which established the first list of prohibited substances. Testing was introduced at the Winter Games in Grenoble and the Summer Games in Mexico of 1968.

Throughout the 1970s, 1980s, and 1990s, doping flourished as new designer drugs and techniques such as blood doping were devised by dishonest athletes and coaches attempting to gain advantage. Even state-sponsored doping practices existed in some countries; the former German Democratic Republic (East Germany) was a major culprit. During this period, technology did not always kept pace with the doping scene; detection was hampered by the lack of reliable, accurate bioanalytical methods.

In the 1990s there was debate in the IOC, other sports federations, and individual governments about the exact definitions of doping, enforcement policies, and punitive sanctions. Confusion continued, and individual sanctions of athletes were often disputed and sometimes overruled in civil courts. As scandals mounted, the IOC convened the First World Conference on Doping in Sport in Lausanne, Switzerland in 1998. Following the proposal of the Conference, WADA was established on November 10, 1999. WADA's role is to promote and coordinate the fight against doping in all sport internationally, that is, to make certain there is a level playing field.

The success of WADA since its inception less than 20 years ago has been mixed. It certainly has not eliminated PED use or cheating athletes; but as an organization, it has made great strides in that direction. Regrettably, there is still much work to be done.

REFERENCES

1. Brooks GA, Fahey TD, White TP. *Ergogenic aids. Exercise Physiology: Human Bioenergetics and Its Application*. 2nd ed. Mountain View, CA: Mayfield Publishing; 1996.

2. Cooper CE, Beneke R. *Drugs and ergogenic aids to improve sport performance. Essays in Biochemistry*. London: Portland Press; 2008.

3. Kelly GS. Nutritional and botanical interventions to assist with the adaptation to stress. *Alternat Med Rev*. 1999;4(4):249–265.

4. Chen CK, Muhamad AS, Ooi FK. Herbs in exercise and sports. *J Physiol Anthropol*. 2012; 8:31–34.

5. Bucci LR. Selected herbals and human exercise performance. *Am J Clin Nutr*. 2000; 72(Suppl. 2):S624–S636.

6. Sandovala M, Okuhamaa NN, Angelesa FM, et al. Antioxidant activity of the cruciferous vegetable Maca (*Lepidium meyenii*). *Food Chem*. 2002;79(2):207–221.

7. Lemon PWR. Is increased dietary protein necessary or beneficial for individuals with a physically active lifestyle? *Nutr Rev*. 1996;54(4 Pt 2):S169–S175.

8. Hackney AC, Dobridge J. Exercise and male hypogonadism: testosterone, the hypothalamic-pituitary-testicular axis and physical exercise. In: Winter S, ed. *Male Hypogonadal Disorders: Contemporary Endocrinology Series*. New York: Humana Publishing; 2003:314–333.

9. Rosenbloom D, Sutton JR. Drugs and exercise. *Med Clin North Am*. 1985;69(1):177–187.

10. *What We Do.* World Anti-Doping Agency. <https://www.wada-ama.org/en/>; Accessed 01.17.2016.

11. Clarkson PM. Nutrition for improved sports performance. Current issues on ergogenic aids. *Sports Med.* 1996;21(6):393–401.

12. Spriet LL. Caffeine and performance. *Int J Sport Nutr.* 1995;5(Suppl):S84–S99.

13. Pittler MH, Ernst E. Dietary supplements for body-weight reduction: a systematic review. *Am J Clin Nutr.* 2004;79:529–536.

14. de Koning Gans JM, Uiterwaal C, van der Schouw Y, et al. Tea and coffee consumption and cardiovascular morbidity and mortality. *Arterioscl Thromb Vasc Biol.* 2010;30:1665–1671.

15. Nelson DL, Cox MM. *Lehninger Principles of Biochemistry.* 6th ed. New York: Macmillian Higher Education; 2013.

16. U.S. Food and Drug Administration. Final rule declaring dietary supplements containing ephedrine alkaloids adulterated because they present an unreasonable risk. *Feder Register.* 2004;69:6787–6854.

17. Jelkmann W, Lundby C. Blood doping and its detection. *Blood.* 2011;118(9):2395–2404.

18. Luck M. *Hormones: A Very Short Introduction.* Oxford: Oxford University Press; 2014.

19. Hackney AC, Kallman A, Hosick KP, Rubin DA, Battaglini CL. Thyroid hormonal responses to intensive interval versus steady-state endurance exercise sessions. *Hormones (Athens).* 2012;11(1):54–60.

20. PDR. *Physicians' Desk Reference.* 70th ed. Montvale, NJ: PDR Network; 2016.

21. Paz-Filho G, Mastronardi CA, Licinio J. Leptin treatment: facts and expectations. *Metabolism.* 2015;64(1):146–156.

22. McMurray RG, Hackney AC. Interactions of metabolic hormones, adipose tissue and exercise. *Sports Med.* 2005;35(5):393–412.

23. *Drug Scheduling.* U.S. Department of Justice, Drug Enforcement Agency. <http://www.dea.gov/index.shtml>; Accessed 01.05.2016

24. American Dietetic Association, Dietitians of Canada, American College of Sports Medicine, Rodriguez NR, Di Marco NM, Langley S. American College of Sports Medicine position stand: nutrition and athletic performance. *Med Sci Sports Exerc.* 2009;41(3):709–731.

25. Clarkson PM, Thompson HS. Antioxidants: what role do they play in physical activity and health? *Am J Clin Nutr.* 2000;72(2):637s–646s.

26. Tandon S, Bowers LD, Fedoruk MN. Treating the elite athlete: anti-doping information for the health professional. *Missouri Med.* 2015;112(2):122–128.

CHAPTER 9

Bottom Line: Why Should I Exercise?

This chapter provides a scientific and practical rationale for the benefits of an exercise program, along with strategies for reducing and avoiding common injuries associated with exercising.

WHAT TO EXPECT WHEN EVERYTHING GOES RIGHT

Regular exercise is associated with many positive adaptations of physiological and biochemical functions. Some leading researchers now consider exercise as the new "polypill" for treating the public health crises of obesity, diabetes, metabolic syndrome, hypertension, and heart disease around the world, a view that is strongly supported by data.[1–3] Nevertheless, people continue to ask why they should exercise. What follows is rationale in straightforward language to answer that question.

Exercise helps control body weight: It prevents excess weight gain and promotes weight loss by creating a negative energy balance. In fact, even if time does not allow a formal workout session, it is still advantageous to be more active throughout the day—taking the stairs instead of the elevator, walking to the post office—can increase energy expenditure.

Exercise combats disease: Research shows that it helps prevent or manage a wide range of health problems, including stroke, hypertension, cardiovascular disease, dyslipidemia (low HDL cholesterol and high triglycerides associated with atherosclerosis), metabolic syndrome, type 2 diabetes, certain types of cancer, arthritis, and falls due to muscular imbalance.

Exercise improves mood: Enhanced self-esteem is a key psychological benefit of regular exercise, possibly through the release of endorphins and other brain neurotransmitters. Endorphins are thought to interact with receptors in the brain to reduce perception of pain and produce positive sensations. The feeling that follows a workout is often described as euphoric and may be accompanied by an energizing outlook on life. The general result is a reduction of stress, reduced anxiety and depression, a boost in confidence, and improved sleep, which is critical for dealing with stress.

Exercise increases overall energy levels: It increases cardiovascular capacity and endurance and improves muscle strength. It increases the delivery of oxygen and nutrients to tissues and helps the cardiovascular-respiratory system work more efficiently, which increases the ability to do work and daily activities without a concomitant need for more rest. There is also increased functional capacity, which reduces the risk of injury.

Exercise improves one's sex life: Chronic fatigue can leave an individual too tired to engage in or enjoy physical intimacy. The exercise-induced physiological and psychological changes noted above, including enhanced self-esteem, alter that situation for the better. Research shows that regular physical activity leads to enhanced arousal in women and reduced erectile dysfunction in men.

Exercise can increase social interaction: Workouts are a means to connect with family, friends, or associates in a pleasurable and different setting. This helps prevent social isolation, which increases the risk of anxiety disorders and depression.

Exercise increases life expectancy: The U.S. Department of Health and Human Services recommends that adults aged 18 to 64 engage in regular aerobic activity for 2.5 h at moderate intensity or 1.25 h at vigorous intensity each week. Agency researchers found that life expectancy was 3.4 years longer for people who reported that they got the recommend level of physical activity (see the Close-Up at the end of this chapter).[1,4]

The bottom line: These benefits, given in this section, can be summarized in the following simple statement:

↑ Regular exercise → ↑ Physiological capacity + ↑ Psychological well-being = ↑ Quality and quantity of life

Table 9.1 summarizes the major physiological and psychological improvements typically found for aerobic and resistance exercise training which lead to an increase in quality and quantity of life.

Table 9.1 Physiological and Psychological Changes Induced by Aerobic and Anaerobic Exercise

Aerobic Cardiovascular Training	Anaerobic Resistance Training
Increased cardiac output	Increased muscle fiber recruitment capacity
Increased blood flow	Improved muscle tissue extensibility
Increased number of capillaries	Increased muscle fiber cross-sectional area
Improved catecholamine responses	Increased joint range of motion
Increased bone mineral content	Increased bone mineral context
Increased mitochondrial density	Increased muscular strength-endurance
Increased energy expenditure	Increased energy expenditure
Increased anabolic hormone production	Increased anabolic hormone production
Increased maximal oxygen uptake	Increased maximal oxygen uptake[a]
Improved movement efficiency	Increased movement efficiency
Improved glucose tolerance	Improved glucose tolerance[a]
Reduced risk of coronary artery disease	Reduced risk of coronary artery disease
Improved blood lipid profile	Improved blood lipid profile[a]
Improved mood (affective state)	Improved mood (affective state)
Reduced risk of depression	Reduced risk of depression
Improved quality of life	Improved quality of life

[a] *Program has significant elements of muscular endurance focus with the training.*

WHAT TO EXPECT WHEN EVERYTHING GOES WRONG

Thousands of people start exercise programs each year, but only a small portion of these individuals remain with it on a regular basis. Regrettably, research suggests that the dropout rate is typically >75%.[5] Many factors contribute to this—lack of time, scheduling problems, lack of access to equipment, waning motivation, weather, social and professional demands, and physical injury. Perhaps the most prevalent is injury. The type of exercise individuals engage in greatly influences the rate of injury, as illustrated in Table 9.2.

Injuries and the Recovery Process

Most people who maintain an active exercise program are likely to be injured to some degree at some time. Following are the most prevalent injuries, with suggested steps for treatment and prevention.[7]

Table 9.2 Number of Injuries Per 1000 h of Activity[6]
Rugby, lacrosse — 30
Basketball, squash — 14
Running (high intensity) — 11
Alpine skiing — 8
Rowing machine — 6
Treadmill (walking or jogging) — 6
Tennis — 5
Dancing classes — 5
Resistance training with weight machines — 4
Resistance training with free weights — 4
Outdoor cycling — 3.5
Stationary cycling — 2
Stair climbing — 2
Walking — 2

Soft-Tissue Injuries

These fall into two general categories, acute and overuse.

Acute injuries are caused by a sudden trauma, such as a twisting action, fall, or a blow. Three major types are sprains, strains, and contusions.

Overuse injuries occur gradually over time, when exercises are repeated too frequently and there is insufficient time to recover between exercise sessions. Common examples are tendinitis and bursitis. An overuse injury typically stems from

training errors, which occur when the exerciser takes on too much activity too quickly—goes too fast, exercises for too long, or does too much of one type of exercise,

technique errors, which result from poor form, especially in resistance training, and overloads muscles too much.

A *sprain* is a stretch or tear of a ligament, the tissue that connects one bone to another. Ligaments stabilize and support joints. The areas most vulnerable to exercise sprains are ankles, knees, and wrists. Sprains are classified by severity:

Grade 1 (mild): Slight stretching and some damage to the fibers of the ligament.

Grade 2 (moderate): Partial tearing of the ligament, with abnormal looseness (laxity) in the joint when it is moved in certain ways.

Grade 3 (severe): Complete tear of the ligament, which causes significant instability and makes the joint nonfunctional.

Table 9.3 The RICE Treatment for Common Acute Soft-Tissue Injuries
• *Rest*—Take a break from the activity that caused the injury. Your doctor may recommend crutches to avoid putting weight on your leg.
• *Ice*—Use cold packs for 20 min several times a day. Do not apply ice directly to the skin.
• *Compression*—To prevent additional swelling and, or blood loss, wear an elastic compression bandage.
• *Elevation*—To reduce swelling, elevate the injury higher than your heart while resting.

Pain, bruising, swelling, and inflammation are common to all three grades although the intensity of each symptom varies with the severity grade of the sprain. Mild sprains are treated by RICE (see Table 9.3 for details) and sometimes physical therapy rehabilitation exercises. Moderate sprains often require a period of bracing. Severe sprains may require surgery.

A *strain* is an injury to a muscle or a tendon, which attaches muscle to bone. Exercise-related strains often occur in the foot, leg (typically the hamstring), or back. Symptoms are similar to sprains and include pain, muscle spasm, muscle weakness, swelling, inflammation, and cramping. Treatment for a mild or moderate strain is the same as for a mild sprain. Surgery may be required for a serious tissue tear.

A *contusion* is a bruise caused by a direct blow or repeated blows that damage underlying muscle fibers and connective tissue without breaking the skin. A contusion can also result from falling or jamming the body against a hard surface. The discoloration is caused by blood pooling around the injury under the skin. Most contusions are mild and respond to RICE treatment.

Tendinitis is inflammation or irritation of a tendon or the covering (sheath) of a tendon. It is caused by a series of small repeated stresses. Symptoms are swelling and pain that worsens with activity. It is treated by rest to eliminate stress, antiinflammatory medication, steroid injection, splinting, and exercises to correct muscle imbalance and improve flexibility. Persistent inflammation may cause significant damage to the tendon and require surgery.

Bursitis is inflammation of the bursae, small fluid filled sacs around the shoulder, elbow, hip, knee, and heel. Positioned between bones and soft tissues, they act as cushions to reduce friction. In many people, bursitis is associated with tendinitis. Repeated small stresses and overuse in exercise cause the bursa to swell. Treatment consists of changes in activity and antiinflammatory medication. If swelling and pain do not respond, a physician can remove fluid from the bursa or inject an antiinflammatory drug directly into it.

Other Injuries

A *myocardial infarction* (heart attack) occurs when blood flow to a part of the heart stops, usually because of a coronary blood vessel occlusion, leading to the death of heart muscle cells. The most common symptom is chest pain (angina) which may travel into the shoulder, arm, back, neck, or jaw (centered or left side) and lasts for more than a few minutes. Other symptoms are shortness of breath, nausea, feeling faint, a cold sweat, and feeling tired. The rate of infarction-induced sudden death in persons exercising is extremely low. Thompson and colleagues, in conjunction with the American Heart Association, reported an annual incidence of 0.75 and 0.13 occurrences per 100,000 young male and female athletes, respectively, and 6 per 100,000 middle-aged men during exertion.[8,9]

Overtraining syndrome (OTS) is a medical condition that occurs when an athlete's stress load becomes excessive, maladaptation rather than positive adaptation occurs, and performance declines.[10] In the field of sports physiology, OTS is viewed as the result of a training plan that is not balanced in the levels of exercise stress load, non-training (life) stress load, and rest. The athlete moves from an appropriate training state to overreaching (OR) and ultimately to the overtraining (OT) state if adjustments are not made.

- OR results from an accumulation of training and/or nontraining stress resulting in a short-term decrement in performance capacity with or without related physiological and psychological signs (see Table 9.4) of maladaptation, in which restoration of performance capacity may take from several days to several weeks.
- OT results from an accumulation of training and/or nontraining stress resulting in a long-term decrement in performance capacity with or without related physiological and psychological signs of maladaptation, in which restoration of performance capacity may take several weeks or months. OTS is a consequence of OT.

The difference here between these two training conditions is the amount of time needed for performance restoration, not the type or duration of training stress or degree of physiological impairment if any.[11]

OR is used by elite-level athletes and their coaches to enhance sporting performance. Such periods of intensified training can result in a transient decline in performance; however, when appropriate periods of recovery

Table 9.4 Pathophysiologic Characteristics in Hypo- and Hyperarousal Forms of Overtraining Syndrome

Parasympathetic, Hypoarousal	Sympathetic, Hyperarousal
Decreased physical performance	Decreased physical performance
Easily fatigued or lethargic	Easily fatigued
Depression	Hyperexcitability
Normal or disturbed sleep	Disturbed sleep
Normal constant weight or weight loss	Weight loss
Low resting HR	Increased resting HR and BP
Hypoglycemia during exercise	Slow recovery of HR and BP after exercise
Loss of competitive desire	Loss of competitive desire
Amenorrhea in women	Amenorrhea in women
Hypogonadism in men	Hypogonadism in men
Increased incidence of infections	Increased incidence of infections
Decreased maximal lactate response to exercise	Decreased maximal lactate response to exercise

Note: *HR, heart rate; BP, blood pressure.*

are provided a supercompensation occurs and performance is greatly enhanced compared to baseline. This short term, effective form of OR is called functional OR (FOR). But if FOR continues too long (ie, weeks) it becomes nonfunctional OR (NFOR), which becomes OT, and the athlete moves toward OTS. These events and the progression can be compounded by inadequate nutrition, illness, and sleep problems.[11]

Distinguishing when NFOR is becoming OT is difficult and depends usually on the diagnosis of a healthcare provider. In most situations this is a retrospective diagnosis, OTS has already developed, and it is too late. Even though the cause of OTS is well known to be an imbalance between stress loads and rest, determining on an individual basis how much stress is too much for an athlete is not easy.

Two varieties of OTS have been proposed, a hypoarousal and a hyperarousal form.[10–12] This categorization is based upon the finding of divergent symptomology in some physiological and psychological parameters of athletes. Hypoarousal is also called parasympathetic or Addison's OTS. It is commonly seen in endurance athletes (long-distance runners, rowers, cross-country skiers, cyclists, and swimmers). Hyperarousal is also called sympathetic or Basedow's OTS. It is commonly seen in power athletes (sprinters, jumpers, and weight lifters) and occurs slightly less frequently than the hypoarousal form. The two forms have some similar characteristics and warning signs, particularly the persistent decline in physical performance. Table 9.4 compares them. Athletes can display some or all of the characteristics noted.

Put simply, OTS reflects the unsuccessful attempt of the body to cope with the physiological and psychological stress of exercise training and life—the total allostatic load and resultant wear and tear on the body from chronic stress.[11,12] Thus, medically it can be understood partly within the context of the general adaptation syndrome of Hans Selye.[13]

Rhabdomyolysis is a condition in which skeletal muscle is damaged and breaks down rapidly. It can be caused by crush injury, strenuous exercise, medications, drug abuse, and infections. Breakdown products are released into the bloodstream, a critical one being the protein myoglobin, which can harm the kidneys and lead to kidney failure. The severity of the symptoms—muscle pain, vomiting, bloody urine, confusion—depends on the extent of muscle damage. Exercise-induced (exertional) rhabdomyolysis is rare. The incidence in the intensive and demanding training done by the US military is 22 per 100,000 per year.[14] The incidence in the general public is much lower but when it occurs, it is serious and requires medical attention.[14]

PREVENTION: TIPS FOR SUCCESS

Injury often happens when people suddenly increase the duration, intensity, or frequency of activities. Many soft-tissue injuries can be prevented through proper conditioning, training, and equipment. Other prevention tips include the following[7,10]:

Use proper equipment: Replace athletic shoes when they wear out. Wear comfortable, loose-fitting clothes that let you move freely and are light enough to release body heat.

Develop a balanced program that incorporates cardiovascular exercise, strength training, and flexibility. Add activities and new exercises carefully. It is best to add no more than one or two per workout session.

Warmup to prepare for exercise: This can be light stretching and some preliminary movement such as walking before the main part of the workout session.

Drink water to prevent heat illnesses—dehydration, heat exhaustion, and heat stroke: Drink 0.5 L 15 min before starting and another 0.5

to 1.0 L after cool down. If possible, drink 0.2 to 0.5 L every 20 min during exercise.

Cool down in the period immediately after the main portion of the workout session, gradually allowing the body to return to a resting state. Cool down should take twice as long as warmup.

Rest by scheduling regular days off from vigorous exercise, and rest when you are tired. Fatigue and pain are good reasons not to exercise.

Avoid "weekend warrior" syndrome: Regular exercise sessions should be spaced throughout the week. Packing all of your exercise into the weekend is ill-advised.

Consult a physician if an injury occurs, whether acute or due to over-use, especially if symptoms do not respond to RICE. Be sure to tell your physician if you have recently made changes in your workout technique, intensity, duration, frequency, or mode of exercise. This will assist in the identification of the cause and the prescription of corrective action.

Close-Up: How Many Years of Life Expectancy Can I Gain by Exercising?

In a landmark study, researchers at the Division of Cancer Epidemiology and Genetics, National Cancer Institute, U.S. National Institutes of Health examined data from nearly 640,000 men and women aged 40 and over for almost 10 years. During this period, about 82,500 died. The researchers looked at lifestyle factors related to these deaths, specifically physical activity levels and body mass index (BMI), a ratio of body mass to height used as an index of proper body weight (elevated BMIs reflect being overweight or obese).[4] Following are the key findings:

- A regular activity level equivalent to brisk walking for up to 75 min/week was associated with a gain of 1.8 years in comparison to sedentary participants.
- Greater activity was associated with greater gain—4.5 years at the highest level, brisk walking for ≥450 min/week.
- Being active (about 150 min/week) and having a normal body weight (BMI < 25) was associated with a gain of 7.2 years relative to being sedentary and extremely obese (BMI > 35).
- Being overweight (but not obese) and doing 150 min/week of moderate activity was associated with a gain of 3.9 years relative to being over-weight and sedentary.
- Even obese participants gained from activity. Being obese and doing 150 min/week of moderate activity was associated with a gain of 2.7 to

3.4 years, depending on the degree of obesity, relative to being obese and sedentary.

This study provided compelling evidence of a clear, direct dose—response relationship between the volume of physical activity and years of life gained. Physical activity, even if you are overweight or obese, can increase your life expectancy. In other words, being physically active pays off.

REFERENCES

1. U.S. Department of Health and Human Services. *2008 Physical Activity Guidelines for Americans*. Atlanta, GA: USDHHS. < http://www.health.gov/paguidelines; 2008 >.

2. Fiuza-Luces C, Garatachea N, Berger NA, Lucia A. Exercise is the real polypill. *Physiology (Bethesda)*. 2013;28(5):330—358.

3. American College of Sports Medicine. *ACSM's Guidelines for Exercise Testing and Prescription*. 9th ed. Philadelphia, PA: Lippincott Williams & Wilkins; 2014.

4. Moore SC, Patel AV, Matthews CE, et al. Leisure time physical activity of moderate to vigorous intensity and mortality: a large pooled cohort analysis. *PLoS Med*. 2012;9(11): e1001335.

5. Garber CE, Blissmer B, Deschenes MR, et al. American College of Sports Medicine position stand: quantity and quality of exercise for developing and maintaining cardiorespiratory, musculoskeletal, and neuromotor fitness in apparently healthy adults—guidance for prescribing exercise. *Med Sci Sports Exerc*. 2011;43(7):1334—1359.

6. Requa RK, DeAvilla LN, Garrick JG. Injuries in recreational adult fitness activities. *Am J Sports Med*. 1993;21(3):461—467.

7. Houglum P. (Athletic Training Education Series) *Therapeutic Exercise for Musculoskeletal Injuries*. 3rd ed. Champaign, IL: Human Kinetics; 2010

8. Thompson PD. The cardiovascular complications of vigorous physical activity. *Arch Intern Med*. 1996;156(20):2297—2302.

9. Thompson PD, Franklin BA, Balady GJ, et al. Exercise and acute cardiovascular events placing the risks into perspective: a scientific statement from the American Heart Association Council on Nutrition, Physical Activity, and Metabolism and the Council on Clinical Cardiology. *Circulation*. 2007;115(17):2358—2368.

10. Hackney AC. Clinical management of immuno-suppression in athletes associated with exercise training: sports medicine considerations. *Acta Med Iran*. 2013;51(11):751—756.

11. Meeusen R, Duclos M, Foster C, et al. Prevention, diagnosis, and treatment of the over-training syndrome: joint consensus statement of the European College of Sport Science and the American College of Sports Medicine. *Med Sci Sports Exerc*. 2013;45(1):186—205.

12. Hackney AC. Stress and the neuroendocrine system: the role of exercise as a stressor and modifier of stress. *Expert Rev Endocrinol Metab*. 2006;1(6):783—792.

13. Selye H. The evolution of the stress concept: stress and cardiovascular disease. *Am J Cardiol*. 1970;26(3):289—299.

14. Alpers JP, Jones LK. Natural history of exertional rhabdomyolysis: a population-based analysis. *Muscle Nerve*. 2010;42(4):487—491.

CHAPTER 10

Future Scientific Research in Exercise and Sport and the Role of Bioanalytical Chemistry

This chapter looks forward to the research questions, advancing technologies, and methods that I think will inform exercise science in the near future—what needs to be studied and how to study it.

Bioanalytical chemistry is a subdiscipline of analytical chemistry involving the separation, detection, identification, and quantification of biological samples in various settings.[1] In exercise science, the objects of study are often molecules such as proteins, peptides, DNA, and drugs, and the objective is to clarify key concepts and answer unresolved questions about how to use exercise to treat and prevent disease and improve physical and mental health.[2–5] The procedures and techniques of bioanalytical chemistry and related fields are critical to unlock the answers about the role and usefulness of exercise. The following are some thoughts on how this quest might proceed in the future.

ANALYTICAL TECHNOLOGY ADVANCES

Although there is no strict "Moore's Law" (that the number of transistors in an integrated circuit doubles about every 2 years) corollary for bioanalytical chemistry, the technology that drives and limits laboratory research is evolving rapidly. Procedures that were inconceivable 20 years ago are now mainstream.[1,6] The gene-chip microarray, for example, allows dissection of gene function on a mass scale. The resulting genetic profiling shows expression patterns in cells and changes thereto induced by disease and physiological or pharmacological intervention. This enables identification of specific sets of genes involved in and being influenced by exercise regimens, nutritional profiles, and environmental conditions (epigenetic factors; see the Close-Up at the end of this chapter) that subsequently affect the human adaptation process. Thus through this line of evolving study insight is and will be gained into the role of exercise in changing the disease processes associated with inactivity and aging.

We are on the cusp of using personalized medicine—the customization of healthcare based on the patient's genetic makeup and epigenetic status—to treat disease, and this model will certainly spill over into exercise prescription and program design by using the developing science of functional genomics.[2,5,7] Furthermore, novel genomic techniques such as next-generation DNA sequencing will find uses in detecting polymorphisms (ie, one of two or more variants of a particular DNA sequence) in populations, which has the potential to impact dramatically on individual differences in the adaptive processes.

The following are a few bioanalytical chemistry tools that are of critical use in this developing and evolving work[1,6,8]:

- The *ASO probe* (allele-specific oligonucleotide) is used to detect the presence of a specific DNA sequence or to visualize a specific polymerase chain reaction (PCR) product after electrophoresis. PCR is a fast and inexpensive technique for copying small segments of DNA so that more extensive analyses can be done.
- The *biochip* is a microarray (a collection of miniaturized test sites) arranged on a solid substrate that permits many simultaneous tests to be performed, allowing higher throughput volume and speed. A variation that is rapidly evolving is the digital microfluidic biochip, which uses a fluid substrate.
- *Digital PCR* is an alternate to conventional quantitative PCR for absolute quantification and rare gene allele detection. The procedure can use nanofluidic chips to run thousands of PCR reactions and analyses.
- *Next-generation sequencing*, also known as high-throughput sequencing, includes such specific methods as illumina sequencing, which uses fluorescent labeling, and ion torrent and ion proton sequencing, which, rather than using an optical signal, measures pH or H^+ responses in analysis of DNA and RNA.

Molecular biology addresses the molecular basis of biological activity between the various systems of a cell, including the interactions between and regulation of DNA, RNA, and proteins and their biosynthesis. It has incorporated a rapidly developing plethora of powerful bioanalytical chemistry techniques for DNA and RNA analysis and is providing new and exciting insights into the proteome, that is, the entire complement of proteins, including the modifications made to a particular set of proteins in an organism or a cellular system.[1,3,8] The young science of proteomics (the comprehensive study of a specific proteome, including information on protein abundances, their variations and modifications, along with their interacting partners and networks, in order to understand cellular processes) will continue to take advantage of rapidly developing analytical techniques.[9] Proteomics is seen as a corollary to genomics, in that it encompasses the functions of all proteins and their interactions. This emerging focus has led to speculation that the study of proteomics will shift major aspects of biological science back toward a more traditional bio-chemical and physiological analytical perspective.[3–5]

FUTURE TRENDS AND RESEARCH PRIORITIES

During the last few years, the US National Institutes of Health has suggested several research priorities that will greatly affect the exercise sciences[2,5]:

- Exploiting genomics by assembling genetic information about the predisposition to diseases, predicting responses to epigenetic factors, and designing new therapies;
- Reinvigorating clinical research with renewed focus on multi-dimensional, integrated clinical trials and translating the findings of basic science to practical application (ie, translational medicine);
- Greater use of the basic sciences of biology, chemistry, and engineering to initiate interdisciplinary programs with medicine to foster innovative drug and molecular development initiatives;
- Improvement of living conditions, lifestyles, and public health knowledge, and dissemination of information to all components of US society as well as other nations, particularly with regard to the aging of the world population.

Exercise scientists, in collaboration with their basic science and medical colleagues, should assume leadership in expanding clinically relevant research in the areas noted above, specifically by partitioning out and targeting the effects of exercise in the etiology and treatment of cardiac, respiratory, hormonal, metabolic, and musculoskeletal disorders.

WHAT ARE THE HEALTH CHALLENGES OF TOMORROW?

Life expectancy in the United States in 1910 was 48.4 and 51.8 years for men and women. In 2010 it was 76.3 and 81.3. For most other countries the trend, if not the magnitude, is similar.[10] Many factors are responsible for this increase, better public health practices and medical care being among the most prominent. But more needs to be done. For example, in recent decades there has been a marked increase in cases of congestive heart failure and other cardiovascular disorders.[5,11] Metabolic diseases associated with obesity and type II diabetes are epidemic.[11] The dramatic increase in longevity brought with it questions about the quality and productivity of extended life, with its increase in age-related morbidities such as osteoporosis, sarcopenia, and frailty, for which a sedentary lifestyle with insufficient aerobic and resistance-loading activities is a clear risk factor. Mitigating these diseases and degenerative processes will require an integrative, multi-disciplinary perspective using the technology evolving from the fields of functional genomics and proteomics as well as basic and applied exercise research. But a large part of the solution to these problems is simple, economical, available to most people, and proven effective: sensible diet, regular exercise, adequate rest, and recreation.

CONCLUSION

Engaging in exercise is for many people in modern society is an option, not an obligation. It is a lifestyle choice. Research evidence has shown it to be a powerful and compellingly helpful lifestyle choice. It has not only short-term but also long-term benefits, and some of those benefits are passed on as traits to offspring (see the Close-Up at the end of this chapter). It has the potential to form us into healthier, more functional, happier people who live longer. That said, there are important details about how to exercise, how much to exercise, how exercise induces it affects, and how it interacts with other lifestyle factors that we just do not know yet. We have questions, and they need to be answered if we are to fully understand how to use exercise in a way that benefits everyone.

Close-Up: Are Great Athletes Born or Made?

Scientists, athletes, and coaches continue to argue over whether superior athletic ability is primarily a product of heredity or training. The new science of epigenetics suggests that perhaps the answer is both.[12]

Epigenetics is the subdiscipline of genetics that studies cellular and physiological phenotypic trait variations caused by external or environmental factors. These factors affect traits by switching genes on and off as well as affecting how cells read genes. This differs from the traditional doctrine of trait variation occurring by permanent changes in the DNA sequence, that is, mutation. Factors such as lifestyle choices (diet, exercise, drug use and abuse), socioeconomic status (financial stress), and environmental exposure (toxins, diseases, climate) are all possible epigenetic influencers.

Mounting evidence indicates that some epigenetic tags remain in place as genetic information is passed from generation to generation: epigenetic inheritance. This goes against the idea that inheritance happens only through a fixed DNA code passing from parent to offspring. It means that a parent's experiences, in the form of epigenetic influenced changes (ie, tags), can be passed to future generations, and there is little doubt that epigenetic inheritance is real. It explains some patterns of inheritance that geneticists have long puzzled over.

Regular exercise training is a powerful effector of genes.[7,12,13] Hence, what parents do to enhance their physiological capacity and health through regular exercise may to some extent be passed along to enhance the capacity and health of their offspring. The degree and extent to which this is possible is a vital subject of research and for some people should be a motivating factor to be more physically active and engage in exercise.

Epigenetic inheritance adds another dimension to the modern picture of evolution. The genome as we have classically thought of it changes slowly, through the processes of random mutation and natural selection, so it takes many generations for a genetic trait to become common in a population. The epigenome, by contrast, can change rapidly in response to signals from the environment. So for the offspring it will convey some experiences of the parents and yet remain flexible as conditions change. In this way the epigenetic inheritance may allow an organism to continually adjust its gene expression to fit its environment, an extremely telling indication of what a highly adaptive organism we humans are.

REFERENCES

1. Gault VA, McClenaghan NH, eds. *Understanding Bioanalytical Chemistry: Principles and Applications.* Hoboken, NJ: Wiley-Blackwell; 2009.

2. Baldwin KM, Haddah F. Research in the exercise sciences: where we are and where do we go from here—Part II. *Exerc Sport Sci Rev.* 2010;38(2):42–50.

3. Booth FW, Laye MJ. The future: genes, physical activity and health. *Acta Physiol (Oxford).* 2010;199(4):549–556.

4. Thyfault JP, Booth FW. Lack of regular physical exercise or too much inactivity. *Curr Opin Clin Nutr Metabolic Care.* 2011;14(4):374–378.

5. Booth FW, Laye MJ, Roberts MD. Lifetime sedentary living accelerates some aspects of secondary aging. *J Appl Physiol.* 2011;111(5):1497–1504.

6. Kuepper S, Haider M. Process analytical chemistry—future trends in industry. *Analyt Bioanalyt Chem.* 2003;376(3):313–315.

7. Kanherkar RR, Bhatia-Dey N, Csoka AB. Epigenetics across the human lifespan. *Front Cell Develop Biol.* 2014;2:49: <http://dx.doi.org/10.3389/fcell.2014.00049>.

8. Hahn MA, Singh AK, Sharma P, Brown SC, Moudgil BM. Nanoparticles as contrast agents for in-vivo bioimaging: current status and future perspectives. *Analyt Bioanalyt Chem.* 2011;399(1):3–27.

9. Fridman E, Pichersky E. Metabolomics, genomics, proteomics, and the identification of enzymes and their substrates and products. *Curr Opin Plant Biol.* 2005;8:242–248.

10. *Life expectancy by age, race, and sex,* 1900–2010. Center for Disease Control and Prevention. <http://www.cdc.gov/nchs/fastats/life-expectancy.htm>; Accessed 12.28.2015.

11. *Chronic disease prevalence.* Department of Health & Human Services, Office of Disease Prevention and Health Promotion. <http://www.healthypeople.gov/2020/about/foundation-health-measures/General-Health-Status>; Accessed 12.29.2015.

12. Espstein D. *The Sports Gene: Inside the Science of Extraordinary Athletic Performance.* New York: Penguin Group; 2013.

13. Hackney AC. Epigenetic aspects of exercise on stress reactivity. *Psychoneuroendocrinology.* 2015;61:17–18.

Note: Page numbers followed by "*f*," "*t*," and "*b*" refer to figures, tables, and boxes, respectively.

F

FAAP. *See* Free amino acid pool (FAAP)
Fad diets, 61*b*
FADH2. *See* Flavin adenine dinucleotide (FADH2)
Fat, defined, 15
Fibers, categories of, 70–71
First law of thermodynamics, 3
Flavin adenine dinucleotide (FADH2), 13–15
fMRI. *See* Functional magnetic resonance imaging (fMRI)
Food intake, 55
   and metabolic rate, 50–51
FOR. *See* Functional overreaching (FOR)
Free amino acid pool (FAAP), 17, 17*f*
Frequency, 65, 66*t*, 74*t*
Functional genomics, 108, 110
Functional magnetic resonance imaging (fMRI), 9
Functional overreaching (FOR), 102–103
Future health challenges, 110

G

Gas chromatography linked to isotope ratio mass spectrometry (GC/IRMS), 21
Gas concentration measurement, 35–37
GC/IRMS. *See* Gas chromatography linked to isotope ratio mass spectrometry (GC/IRMS)
Gender, and resting metabolic rate, 49
Gene-chip microarray, 107
General adaptation syndrome, 104
Genomics, functional, 108, 110
Geography and climate
   influences of, on energy expenditure, 60–63
Gerschler, Woldemar, 73
Ghrelin, 90
Global questionnaires, 38–39
Glucagon and energy metabolism, 26*t*
Gluconeogenesis pathway, 19
Glucose cycles, 19
Glucose
   as an energy source for ATP, 19
   metabolism of nutrients into, 11–13
Glycogenolysis, 25*t*, 27
Glycolysis, 13–15, 76–77
   key enzymes in, 76
Growth hormone and energy metabolism, 26*t*

H

Häkkinen, Keijo, 80–81
Haldane transformation, 34
Harris–Benedict equations, 37, 38*t*

Health challenges, in future, 110
Health problems, managing, 97
Heart attack. *See* Myocardial infarction
Heart rate, 68
Heart rate monitors, 39
Heart rate reserve (HRR) method, 68, 69*t*
Heat exhaustion, 79–81, 104–105
Heat stroke, 79–81, 104–105
Heavy resistance training, 80
Hemoglobin, 68, 88, 92–93
Herschel, Sir William, 41
Higher-intensity exercise, 58–60
High-intensity interval training (HIIT), 73
High-throughput sequencing. *See* Next-generation sequencing
HIIT. *See* High-intensity interval training (HIIT)
Homeostasis, 25, 25*t*, 79
Hormones, 5, 25–27
Hyperarousal, 103–104
Hypoarousal, 103–104
Hypothalamus, 24–25, 61–63, 90–91

I

Illumina sequencing, 108
Indirect calorimetry, 20–21, 21*f*, 34–37
   gas concentration measurement, 35–37
   oxygen uptake, 35
Infomercials, 61*b*
Infrared absorption spectroscopy, 36
Infrared rays/infrared radiation, 41
Infrared thermography, in exercise energy expenditure research, 41*b*
Injuries and recovery process, 99–104
   myocardial infarction, 102
   nonfunctional OR (NFOR), 103
   overreaching (OR), 102–103
   overtraining (OT), 102
   overtraining syndrome (OTS), 102–104
   prevention, 104–106
   rhabdomyolysis, 104
   soft-tissue injuries, 100–101
Insulin and energy metabolism, 26*t*
Intensity, 6*f*, 51*t*, 65, 66*t*, 68
Internal milieu, 27
IRMS. *See* Isotope ratio mass spectrometry (IRMS)
Isotope ratio mass spectrometry (IRMS), 37

K

Krebs, Sir Hans, 13
Krebs cycle, 13–15, 18, 18*f*, 27, 68
   amino acid metabolism and, 18, 18*f*

Printed in the United States
By Bookmasters